The 7th (Qu _____, Hussars

The 7th (Queen's Own) Hussars

As Dragoons During the
Flanders Campaign, War of the
Austrian Succession and
the Seven Years War

Volume 1: 1688-1792

C. R. B. Barrett

LEONAUR

The 7th (Queen's Own) Hussars: as Dragoons During the Flanders Campaign,
War of the Austrian Succession and the Seven Years War
Volume 1: 1688-1792
by C. R. B. Barrett

Originally published in 1914 under the title
The 7th (Queen's Own) Hussars

Published by Leonaur Ltd

ISBN: 978-1-84677-467-6 (hardcover)
ISBN: 978-1-84677-465-2 (softcover)

http://www.leonaur.com

Publisher's Notes

The opinions expressed in this book are those of the author
and are not necessarily those of the publisher.

Contents

Preface

In the four volumes of this history I have endeavoured to write the history of the distinguished cavalry regiment that in 1914 we know as the 7th (Queen's Own) Hussars. The first volume contains the military history of the regiment from its incorporation down to the conclusion of the pre-Napoleonic period. The second volume continues the story of the regiment through the Napoleonic period. The third volume covers the rest of the 19th century, and the fourth and final volume is concerned with dress, equipment officer rolls and so on. Illustrations in monochrome and line are scattered through the pages of all volumes, and a number of maps and plans also provided. Appendices referred to in all volumes appear at the end of volume four. Some illustrations that appear in the first three volumes appear also in volume four.

The military history begins with the origin of the regiment, and narrates the circumstances which attended the raising of certain Independent troops of horse in Scotland in 1688, which two years later were constituted a complete regiment of cavalry. I recount, as far as I have been able to trace them, the services of the regiment in the various campaigns in Flanders and Germany during the years 1694-1697, 1711-1713, 1742-1749, 1760-1763, 1793-1795 and 1799. During this period the regiment fought many times and oft. The pitched battles in which they were engaged included the historic fields of Dettingen, Fontenoy, Roucoux, Val, Warburg, Vellinghausen or Kirchdenkern, Beaumont and Willems. Into the long list of minor affairs, combats, sieges and skirmishes I need not enter here.

But besides these campaigns the regiment saw active service in the Jacobite Rebellion of 1715. A Light Troop was raised in 1755 and was employed in the two expeditions to the French coast.

In 1783 the 7th (Queen's Own) Dragoons were converted into light dragoons and in 1807 into hussars.

In 1808 they participated in the rigours and hardships of the Coruña Campaign. Five years later they were for the second time sent to the Peninsula, where, after the invasion of France, they fought at Orthes and Toulouse.

On the declaration of peace the regiment returned home. The escape of Napoleon from Elba again called the 7th (Queen's Own) Hussars into the field. Their services were engaged both at Genappe and at Waterloo, on both of which memorable occasions their casualties were heavy.

For reasons, to ascertain which I must refer the reader to the history itself, it has been needful to enter very fully into the events which took place at Genappe on June 17, 1815. The regiment remained in France as a part of the Army of Occupation until 1818. For twenty years the 7th Hussars were not actively engaged, but in 1838 the Service Troops were ordered to Canada, where they were employed against the insurgents in the Lower Province, and one troop was similarly engaged in the following year. It did not fall to the luck of the regiment to be ordered to the Crimea, but in 1857 for the first time the 7th Hussars were despatched to India. Here they were engaged in the strenuous Mutiny campaign, being most actively employed at Lucknow and on various expeditionary services. Order being restored in India, the regiment remained in cantonments until 1863, when they had an opportunity of taking their share in one of our so-called little wars.

As part of a column under the command of Colonel Alexander Macdonnell, C. B. (Rifle Brigade), a squadron of the regiment charged and overthrew a large body of fanatical tribesmen at Shabkadr. This affair is noteworthy as being the only occasion on which British cavalry had an opportunity of distinguishing themselves as a body in a charge against hill-men.

In 1881 the regiment went to Natal, and during the next year a detachment was for a short time stationed in Cyprus.

When in September 1884 a Camel Corps (Heavy and Light) was formed for temporary service in Egypt, the 7th Hussars furnished a contingent to the Light Camel Corps. Their experience of desert warfare necessarily entailed discomfort, though happily the actual war casualties were not heavy. In 1886 the regiment was ordered to India for the second time and remained there until 1895. Next came the Matabeleland and Mashonaland campaigns, and then after three years' service at home the 7th (Queen's Own) Hussars, during 1901 and 1902, took part in the South African War. From 1905 to 1911 the regiment was at home, after which a third period of service in India rather unexpectedly began.

The 7th (Queen's Own) Hussars are now stationed at Bangalore.

This then is a brief résumé of the military history of the regiment. For the rest I have endeavoured to give many details of events, whether at home or abroad, with which the regiment has been concerned. Anecdotes, public occurrences, riots, &c, have been mentioned as far as possible in their chronological place. Augmentations and reductions are duly chronicled, but besides these I have here and there added passages illustrative of the conditions of military life in Great Britain at various dates, and I venture to hope that these may prove not uninteresting to the reader, nay, perhaps even of some little utility to those engaged in military research. I would particularly draw attention to the interesting and valuable diary and correspondence of Lieutenant and Adjutant Hunt, which, being for the first time published, throws considerable new light upon the unfortunate expedition to the Helder in 1799. In this we get the personal note which is so much needed in regimental histories. It is to be regretted that more old diaries and a greater number of letters belonging to former officers of the 7th Hussars are not to be discovered. In this connection I must confess that the writing of this book has been no easy task, and also that it has occupied a somewhat protracted period of time—nearly three years. But the manuscript materials were very scanty, and in consequence the research required was materially increased. The manuscript record of the services of the regiment is by no means voluminous, nor indeed is it the original copy. Probably the original was either lost during the retreat to Coruña or perished in the wreck of the *Despatch* transport in 1809. That a copy existed when Cannon wrote his history is known, for it was sent to the Horse Guards in 1834 and was never returned to the regiment. The present copy, which verbally agrees to a large extent with Cannon, purports to begin in 1690; but as the handwriting is the same down to later than

1840 one is inclined to suppose that the age of the writer was somewhat patriarchal. I have therefore been compelled to turn to other sources for information, and have searched every document or printed book which I thought likely to assist me. How many these may number I cannot say, nor do I intend to fill pages with a list of them. But I wish here to express my obligations to all those from whose labours in the past I have derived assistance.

The chapter on uniform will, I trust, prove interesting, and I have endeavoured to make it as complete as space would permit. The uniform distinction between dragoons and Royal dragoons and similarly that between horse and Royal horse is, I believe, a new find.

The catalogue of the arms used by the regiment since 1690 is, I trust, complete, and the notes on guidons, the band, regimental medals, and regimental sobriquets contain all that I have been able to gather. The curiosities culled from the 18th century inspection reports are sometimes amusing. The handsome collection of regimental plate has a chapter to itself, and I am happy to have been able for the first time to trace the donors of a fine series of antique spoons and forks. These I traced through their bearing not only the regimental badge but also the private crests of the givers. Lists of the chief successes of the regiment on the polo ground and on the steeplechase course complete the work.

Appendices contain alphabetical lists of all the officers, with their commission dates, services, honours, rewards and such biographical details in brief as may be required. They are arranged in periods 1690-1713, 1714-1871 (the abolition of purchase), and 1871-1914..

It may be interesting not only to British but also to American readers to note the entry in Appendix 2 of the name George Washington which may possibly prove to be a fresh fact with regard to the life of that distinguished soldier and statesman.

As regards the illustrations, I regret that there are not more portraits of 18th century officers, but it would appear that such portraits do not exist; at any rate, if existing, their whereabouts could not be traced.

The incidents which Mr. Harry Payne has illustrated appear to me to be most excellent, and they will, I trust, satisfy the reader. The minute attention given by this artist to details of bygone uniform and equipment merits great commendation, and in execution the groups are full of spirit.

To Mr. P. W. Reynolds I am also much indebted for the drawings in colour strictly illustrative of types of uniform at different periods. To Mr. D. Hastings Irwin, and also to Mr. Reynolds, I must express my obligations for much kindly information on doubtful points connected with old time uniforms and also for the tracings of certain sketches in their collections. As is well known, both gentlemen are high authorities on this subject.

Interspersed throughout the narrative will be found drawings of the medals, &c, awarded for the various campaigns in which the regiment has taken part. Official rewards are also included, and likewise the private regimental medals of old time.

I have also to express my great gratitude to Lieut.-Colonel Sir Arthur Leetham for the more than kind interest he has taken in my work and the the great assistance he has given me, not only in his capacity as curator of the museum of the Royal United Service Institution, but also as the honorary managing editor of *The Cavalry Journal*. From the numbers of that periodical indeed, not a few illustrations have been most generously placed at my disposal for this book.

From the library of the institution I have received most invaluable assistance. The entire book was in fact written on the premise I must here express my obligations and thanks to the librarian, Major C. H. Wylly (late 1st Battalion South Staffordshire Regiment), and also to Mr. Harper, the clerk (late 5th Dragoon Guards), whose assistance, not only in questions referred to them concerning books, but also in the selection of maps, has been most freely given to me. Without the use of the library at the institution I will not say that a regimental history could not possibly be written, but it would be almost impossible to write one satisfactorily. I have also had need many times to consult the War Office library, where from the parliamentary librarian, A. D. Cary, Esq., I have, as ever, received most valuable and kindly assistance.

The encouragement and aid which I have experienced from the Colonel of the regiment, Major-General Sir Hugh McCalmont, K.C.B., C.V.O., has been very great, not only in illustrations, but also in the considerable trouble which he has taken to answer numerous queries and to place me in communication with those likely to be able to assist me in my work.

In the production of a book of this description much depends upon co-operation between the printer and the author. The extreme care which Mr. K. R. Wilson (of Messrs. Spottiswoode & Co., Ltd.) has bestowed in aiding me with his technical knowledge, and the many most useful suggestions which he has made, I here most gratefully acknowledge.

Finally, in presenting to the 7th (Queen's Own) Hussars the history of their Regiment, and to such other readers as may be interested in the military history of our country, I trust that despite deficiencies which must exist in the book, it may yet prove on the whole not unworthy of the distinguished corps whose record its pages contain.

C. R. B. Barrett
Royal United Service Institution
April 25, 1914

Author's Acknowledgements

The author desires to express his most sincere thanks to the following ladies and gentlemen who have so kindly permitted him to use for the purpose of these volumes, miniatures, pictures, photographs, prints, letters, &c., and also for the verbal or written information upon many points with which they have been so good as to furnish him:—

The Lady Bateman, Mrs. Bland, Mrs. Fyers, Miss B. Wylie, Major His Royal Highness Prince Arthur of Connaught, K.G., G.C.V.O., A.D.C.; Major His Serene Highness Prince Alexander of Teck, G.C.B., G.C.V.O., D.S.O.; Major-General Sir H. Augustus Bushman, K.C.B. (Colonel, 9th (Q.R.) Lancers); Colonel Harold Paget, C.B., D.S.O.; Colonel Maisey, Lieut.-Colonel John Vaughan, D.S.O.; Lieut.-Colonel R, M. Poore, D.S.O.; Lieut.-Colonel The Right Honourable The Earl Waldegrave, V.D.; Lieut.-Colonel E. W. N. Pedder, Major Robert Poore, Major C. I G. Norton, Major J. Fryer, Major M. Barne, Captain W. Paget-Tomlinson, Captain H. M. McCance, The Rev. W. A. Diggens, R. P. Copeland, Esq., *The Illustrated London News.*

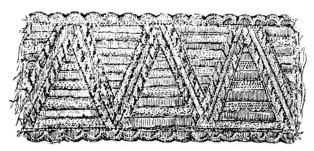

THE REGIMENTAL LACE

13

CHAPTER 1

The Origin of the Regiment
1690-1695

The Regiment known as the 7th (Queen's Own) Hussars originated under circumstances which will be related hereafter. As a preliminary, however, it may be well to mention the following facts.

The Manuscript Regimental Record ostensibly begins in 1690, but internal evidence shows that its pages must have been written in the earlier portion of the nineteenth century, and may even date in part as late as 1840. The account therein is also most fragmentary. Whether Cannon's official History of the Regiment was compiled from records now destroyed, but which were than in existence, cannot well be determined, though this was probably the case, and for this reason.

The MS. Regimental Record contains several details as to routes and expenses while on the march which are evidently authentic, but which are not to be found in Cannon's History.

Unfortunately, the original manuscripts of Cannon's histories, now preserved in the War Office Library, only concern infantry regiments; those of the cavalry of the line are no longer to be found there, if indeed they were ever preserved. However, although many details which would have been full of interest are lacking, it is possible from Cannon and from other sources, as well as from the existing Manuscript Regimental Record, to piece together a fairly accurate and detailed account of the early history of the 7th (Queen's Own) Hussars.

One fact is curious, and that is, that there is a very scanty list of the names of the original officers. In Appendix 1 (volume 4), however, the names of all it has been possible to discover will be found, and this list comprises those who served in the Regiment from 30 December 1690 till its disbandment in Ireland in the spring of 1714.

And here it is also strange that the exact date of this disbandment is unknown. Without entering upon an account of the political situation which arose after the flight of James II, we will now proceed to the account of the origin of the Regiment. To counteract the endeavours of the Jacobites in Scotland who desired the restoration of the exiled King, during the year 1688 three troops of horse were raised, and were commanded by the Earl of Annandale, Lord Belhaven, and William, Laird of Blair, respectively. Sundry independent troops of dragoons were also raised, and besides these several regiments of foot.

It may be remarked that the troops of the Earl of Annandale and Lord Belhaven fought at Killicrankie, 27 July 1689. They were the only cavalry present. The Laird of Blair had previously been captured by Viscount Dundee, and what had become of his troop is not known. There are good reasons, however, for concluding that it had not at that time been properly armed and equipped, and that it melted away, or remained, as it were, disembodied, after the capture of its nominal commander. The Laird of Blair, it may be mentioned, was imprisoned by Dundee in the Isle of Mull, after being carted about the country as a captive. There he died owing to the ill-treatment he had received.

His lieutenant, the Laird of Pollock, who was also a prisoner, likewise suffered much at the hands of his captors.

It will be remembered that Dundee fell during the action at Killicrankie. Major-General Mackay, who commanded the Royalist Army, came off second best in the battle, but managed to conduct the retreat of his shattered troops with some skill; and later succeeded in raising a fresh force. The Jacobites after the loss of their leader withdrew to the mountains, where they remained in temporary security.

Early in the following year (1690) the troops of horse, now again three in number, were formed into a regiment of which the following were the three chief officers: The Earl of Eglington, Colonel; The Hon. William Forbes (Master of Forbes), Lieut.-Colonel and Sir George Gordon, of Edinglassie, Major.

The three troops of dragoons were similarly enrolled as a regiment with Lord Cardross, Colonel (Henry 3rd Lord); Robert Jackson, Lieut.-Colonel and David (Master of Cardross), Captain of a troop and apparently the major's troop. Cannon gives the name of —— Guthrie as the Major, but this requires investigation.

At the same time a regiment of foot was incorporated, and the command thereof was given to Richard Cunningham, who in the following year became the first colonel of the present 7th (Queen's Own) Hussars. For Colonel Cunningham's dates and services see Appendix 1 (volume 4).

A document exists which refers to this and is as follows:

18th December, 1689
Holland House.
Instructions to the Earl of Leven, Major-General Mackay and Sir G. Munroe concerning the forces, You are to form a regiment to be commanded by Richard Cunningham.

(This was a regiment of foot.)

You are to model 3 troops of dragoons, each troop to consist of 50 men and the Lord Cardross to be Colonel and Captain of one troop, Robert Jackson to be Lieut.-Colonel and Captain of the 2nd and Patrick Hume of Polwarth to be Captain of the 3rd, and —— Guthrie (his name was John) to be Major without any troop.

You are to appoint lieutenants and other inferior officers for the said troops of dragoons. You are to complete such regiments first as you think most proper for our service and to transmit a list of such officers as you nominate to George Lord Melville our Secretary for commissions.

It was from these three troops of dragoons that the Regiment eventually sprang. The names of the officers are given, but do not concern this history.

These regiments were actively employed in Scotland during the major portion of the year 1690. The operations of the Royalist Army had been successful; the Jacobites, after being continuously repulsed and harassed, found garrisons strongly posted in the Highland districts. Resistance for the time being gradually ceased, as it was clearly perceived by the disaffected clansmen that active operations would be absolutely futile.

King William now determined on remodelling the Scots army.

Certain of the infantry regiments were broken, but Cunningham's foot was not. Cannon states that:

> . . . . the two regiments of cavalry, of three troops each, were incorporated and constituted a regiment of dragoons, of six troops of fifty men each, of which Robert Cunningham was appointed colonel by commission dated 30th of December, 1690.

Cannon is inaccurate. Cunningham's Christian name was Richard and not Robert, and Lord Cardross's Dragoons certainly numbered six troops. A complete list of the officers is recorded in the Muster Rolls for November and December 1690. Here the name of John Guthrie occurs as Major, but not in command of a troop.

Also Robert Jackson appears not only as Captain of the lieut.-colonel's troop, i.e. lieutenant-colonel, but also beneath Guthrie's name as aide-major. Certain it is that there were six troops in Cardross's Dragoons, but how many men there were in each troop we do not know. But assuming that there were only three troops in the Earl of Eglintoun's horse, we have nine troops out of which the new cavalry regiment was to be formed.

What we do know is this, that by a commission dated 30 December 1690, Richard Cunningham was appointed Colonel of this new regiment, being succeeded in the command of his old regiment of foot by John Buchan. At the same time the Master of Forbes was appointed Lieut.-Colonel, and the Hon. Patrick Hume the Major.

Lastly, the strength of each troop was fifty men each.

Of the other officers who were appointed to the Regiment or commissioned on 30 December 1690 not another name has come down to us; the next commission dates being in June 1691.

Dated from Dublin Cattle, 1 February 1692, the Lords Justices writ to the Earl of Nottingham with regard to supplying horses, as follows:

> We have now in possession about 400 horses, fit for troopers and dragoons, for which we paid but one third in hand and gave a note for the rest. The troop horses were rated according to value, but not exceeding 10s., and the dragoons the like, but not above 5s. We design to deliver them as recruits to the Regiments of Schaack, Oxford, Levenson, Cunning-

ham and Wynne, the commanding officers being willing to take them as such; what remains we shall so dispose of as shall be most to the King's advantage.

This is interesting for several reasons. It shows that Irish horses were at the time bought for the British service. It gives some indication as to price, and lastly tells us definitely that the class of horse purchased for dragoons differed from that bought for Horse.

The 7th (Queen's Own) Hussars, to give it its present title, was therefore originally a Scots regiment. In Scotland Cunningham's Dragoons remained until February 1693 4, their occupation being at first to hold in check the disaffected in the Highlands; and later to take up quarters near Edinburgh.

The date of this movement is thus fixed: a paper is extant, dated 26 March 1692, ordering Colonel Cunningham's Regiment, then in the county of Fife to proceed to Leith (S. P. Dom. King William's Chest, vol. 12, no. 57). The paper is a long one and interesting. For our purpose it is important as fixing the whereabouts of the Regiment at that particular date, and it corroborates the Regimental Records and Cannon's History.

26 March 1692—An interesting light on the military situation in Scotland at this date is to be found in a letter from Lord Melville addressed to the King. The letter (a very long one) contains the following extract and is dated from Edinburgh:

I am just now informed of a tumultuous kind of business which has fallen out here, which may make a noise at a distance, which was this. Colonel Cunningham's regiment was brought from the country of Fife within these few days to the town of Leith, within a mile of this, because there were but few forces about the city, which was not safe to be in time of Parliament, whatever should fall out. There being no pay to give to the soldiers, and the parliament being adjourned, so no expectation of money soon, had it seemed instigated a multitude of women to come here and infest and threaten Major-General Mackay; more of this kind is threatened and feared. The poor people are not able to give the soldiers subsistence, for many have difficulty to subsist themselves.

The letter continues in speaking of the general state of the country, and states that there is some endeavour to 'possess the people with the blackest things of your Majesty that hell can invent.' Unfortunately we cannot learn how the matter ended, but it is clear that, at any rate for the time, the Regiment was in a by no means comfortable station.

By this time proclamations proffering indemnity and pardon to rebels who were prepared to submit and take the oath of allegiance had produced a good effect in the way of pacification. Within six months, that is by the end of January 1692, the clans had submitted, and for a few years Scotland was to be free from war. England was not, however, at peace. William III had gone to war with France. He had been absent from England in order to carry on his campaign, and did not return till the end of 1693. It must be remembered that in those days, as in the days of Caesar, armies were wont to ' go into winter quarters,' and into winter quarters the allied armies went, while William returned to look after his British Dominions.

Early in 1693-4 orders were received by Cunningham's Dragoons to march into England, and then to embark for Flanders in preparation for the campaign of 1694.

The document concerning this is as follows:

Whitehall
9th Feb. 1693/4
Sir, His Majesty having ordered Col. Cunningham's Regiment of Dragoons to March forthwith out of Scotland in order to embark in the river of Thames, for Flanders, is pleased to direct that care be taken for supplying that Regiment with Subsistence upon the English Establishment as soon as they shall arrive al Berwick.
I am, Sir,
Yours &c.
William Blathwayt
Secretary at War
To Mr Guy

Nine days after the date of this letter the Regiment began its march southwards. On arrival at Berwick, Cunningham's Dragoons were duly placed on the English Establishment. The Regimental Record gives the following details as to routes and expenses after leaving Berwick, showing that the destination of each of the six troops was different in the first instance, and that it was not until 21 May that the regiment was fairly embarked at Greenwich.

Troop	From	To	Miles at 8d per mile	Expenses
1	Berwick	Cambridge	236	£7 17 4
1	Berwick	Saffron Walden	245	£8 3 4
1	Berwick	Bishop Stortford	252	£8 8 0
1	Berwick	Newmarket	245	£8 3 4
1	Berwick	Sudbury	261	£8 14 0
1	Berwick	Bury St. Edmund	255	£8 10 0

Fire and candles in the above march and Quarters from 17th Feb. to 10th April. 52 days at 1/- pr day each quarter and 2/- for Headquarters £18 4 0

The expenses for the remainder of the march are not given, but the six troops then converged on London duly arrived there, and were quartered in Holborn and the neighbourhood till 9 May. The wagon expenses of this latter portion of the march are given as follows:

Wagons	Troop	From	Miles	Expenses
1	1	Saffron Walden	40	£1 6 8
1	1	Bury St. Edmund	64	£2 2 8
1	1	Cambridge	45	£1 10 0
1	1	Newmarket	52	£1 14 8
1	1	Bishop Stortford	30	£1 0 0
1	1	Sudbury	50	£1 13 4

Fire and Candles in their march and quarters from the 10th April till the 9th May £10 3 0

Meanwhile officers left behind in Scotland for the purpose had been actively engaged in recruiting. Two entire troops, of sixty men each, were newly raised, and besides these an augmentation of ten men to each of the existing troops had been accomplished. The regimental tradition is that the requisite men were enlisted at and near Hamilton.

30 March 1694—A warrant was issued from Whitehall reciting that 'Two troops of dragoons are to be added to the regiments of dragoons commanded by Major General Sir Thomas Levingston and Colonel Cunningham respectively, consisting each of two ser-

jeants, two drummers and sixty private soldiers. An addition of ten men is also to be made to each of the six old troops and the necessary arms are to be provided for arming the same.'

The two new troops were marched with speed to London, via Huntingdon, in April. We read:

Wagons	From	To	Miles at 8d per mile	Expenses
2	Berwick	Huntingdon	224	£14 8 8
Fire and candle on their march and quarters from 7th April to 20th May				£6 12 0
2	Huntingdon	London		£3 6 8
Fire, candles and quarters				£1 10 0

From the following paper it is clear that considerable exertions must have been made to thus augment the Regiment and to get the new troops to London:

Hague
8/18 May 1694
Sir, His Majesty is pleased to order that in case the two additional troops of Col Cunningham's Regiment of Dragoons do not arrive in time so as to embark with the Regiment they do then embark with the six troops of Col Wynne's Regiment, and ye four other additional troops. Wherein you will take care that they receive the necessary orders. I refar you for our news to ye enclosed paper (missing) and am, your most humble servant
W Blathwayt
Mr. Clark

18 May 1694—A warrant was issued to Henry, Earl of Romney, directing that out of the Ordnance Stores 'one hundred and twenty pair of pistols are to be issued for arming the two additional Troops in the Regiment of dragoons commanded by Colonel Cunningham.' This is interesting as proving that the Regiment was armed with pistols, a pair to each private man at that date.

They did, however, arrive in time, and the whole eight troops embarked at Greenwich on 20 May 1694. Even in these early days, therefore, the zeal for His Majesty's service was a marked feature in the conduct of Cunningham's Dragoons.

The charges on 21st May for wagons, fire, candles and embarkation expenses are riven as follows:

Wagons	Troops	To	Miles	Expenses
2	6	Greenwich	6	£1 10 0
2	2	Greenwich	6	8 0
Fire and candles at Greenwich 2 nights				18 0
Extraordinary charges for embarking				£7 10 0

24 May 1694—A news-letter from London addressed to the Earl of Derwentwater at Newcastle-on-Tyne contains the following extract:

Tomorrow, six additional troops of dragoons for the Regiments of Essex, Fairfax and Cunningham, embark for Flanders, and on June 6 the remaining six troops of Wynne's dragoons embark for the same, which will be the last of the troops that will be sent thither for the service of this campaign.

It should be observed that there was another Colonel Cunningham at that time in command of a regiment of dragoons. With regard to this officer and his regiment there are many entries in the State Papers, and much care is needed in order to avoid confusing one with the other.

This regiment of dragoons (not the 7th) was raised and served in Ireland. It was at or near Derry in 1689, and appears to have returned thence to Dublin, reporting that 'the town was in hazard of surrendering'.

On 24 May of that year its Colonel (Cunningham) was committed to the Gatehouse. Three troops of the regiment were brought to England in 1794 from Dublin, it being proposed to send them to Flanders, an intention which was not carried out. The voyage appears to have been prosperous, at least no misadventures are recorded.

The Regiment arrived at its destination, Willemstadt (Williamstadt), on 31 May. Willemstad, a town in N. Brabant, is situated on the middle mouth of the delta of the river Maas (Meuse). It is about 33½ miles due north of Anvers (Antwerp), and about 16 miles south-east of Helvoetsluys (Sluys), though upon the opposite bank of the estuary. King William was at this time at Breda, a town about 15 miles south-east of Willemstad; this we learn from the following document:

Breda

22 May 1694

Sir, This serves to accompany the enclosed orders for your march towards Brabant with the Regiment under your command which orders it is H. Majesty's pleasure you observe notwithstanding the former received by you for the marching into Flanders.

I am Sir, yours &c.

To the Officer in Chief with ye Regiments of Dragoons commanded by Sir Thomas Levingston & Colonel Cunningham, at or about Willemstadt.

The enclosed orders, however, are not forthcoming, neither are the former orders apparently in existence.

The regiments of dragoons duly left Willemstad, though the precise date of departure is not known. However, on 16 June the King reviewed Sir Thomas Levingston's and Colonel Cunningham's Dragoons that had lately come over from Scotland and were cantoned near the town of Aerschot. The review actually took place at Tremont. The main army was now in camp at Mont St. André near Iodoigne. Its position is thus stated, though it does not seem possible to identify on modern maps the villages named.

The right was at Harlue and Taviers on the Mehaigne. The left was beyond Marilles and in rear of Molembaix. The line formed a sort of elbow near the village of Henieux Heddin, which was in front. The right of this camp was a plain, and on the left it went to narrow and close grounds.

Here at Mont St. André all the cavalry joined the main army.

All the dragoons, commanded by Major-General Eppinger and consisting of twenty-six squadrons of Dutch and twenty-eight squadrons if British, were now stationed to form the reserve.

The brigades of British were thus composed :

	Squadrons		Squadrons
Brigadier Wynne		Brigadier Matthews	
Eppinger	5	Fairfax	4
Essex	4	Levingston	4
Wynne	3	Mathews	4
Cunningham	4		

24

This was a force of twenty-eight squadrons of British cavalry which amounted to about 5400 men.

On 22 July all the dragoons were sent to the rear to a place named Hottemont.

On. 8 August 1694 the army decamped from Mont St. André to march to Rosselaer (or Rousselaer).

Two days previously, to prevent disorders on the march and the probable cutting off of small parties by the enemy, an order was issued to be read at the head of every regiment. In it these offences were forbidden, any breach being punishable by death.

1. Maroding (Marauding).
2. Foraging without orders.
3. Molestation of victuallers or any persons who came to camp with provisions.

The army marched towards the spring of the Mehaigne and Gemblours (Gembloux), passing by the defile of Perwys.

The King took up his quarters at the Château of Sombref, where, being in front of the army, he was covered by all the English and Dutch dragoons, who encamped in a long line.

The French, who were near the plains of Fleury, however, declined battle, and retired over the Sambre.

The allied army then marched to Nivelle, via Mellé and Genappe. The right rested on Arkennes, the left extended to near Promel.

Here a camp was formed for a few days.

On this portion of the march a suspicious-looking stranger was observed in conversation with the driver of one of the ammunition carts. He was seized, and on being searched a lighted match was discovered concealed on his person. It was found on inquiry that he had ingratiated himself with the driver by means of drink money. Persuasion was applied to the prisoner, who then confessed that his intention had been to blow up the cart. Tried by court-martial, he was condemned, and was subsequently put to death. The manner of his execution does not, however, commend itself to modern ideas of punishment. A fire was lighted, and the man's right hand was first cut off and then burnt before his eyes. That being accomplished, the rest of him was cremated alive. It appears that in the previous campaign a similar case had occurred, and that the punishment was precisely the same.

On 12 August the army marched to Aeth, crossing the Cambron at Lens. The King's quarters were at Chievres. Aeth, a town at the confluence of the rivers Cambron and Dender, was finely fortified by Vauban. Edward d'Auvergne gives a long and interesting description of the place and its three fine gates.

The next camp was at Grames, a place about four miles from Tournai and close by Mount Trinity. King William took up his quarters at the Château de Cordes, while the Elector rested at Chastelet. Meanwhile the French were at Courtrai. The King passed the Scheldt at a spot two leagues below Oudenarde. The French had made a most terrific march as regards speed, to cut him off, but failed. It is stated that their losses in men and horses on this hurried march were terrible, and that the stench of dead men and animals tainted the air for miles along the road.

At Escanass King William came in sight of the French army, but no battle ensued. The next halting-place of William was Berghem.

On 7 September all the English horse and dragoons were sent to Rousselaer from Wouterghen. Three days later the King went toward Dixmuyden. Here he reviewed the English horse and dragoons, finding all in good order.

But although there had been no actual fighting in this campaign, in regard to marches it had been very arduous and harassing. A paper issued about this time is of interest and its contents may be summarised.

> From Camp at Rosselaer
> 19/29 Sept. 1694.
> Horse and dragoons are to be completed as regards men, horses, clothes and accoutrements before 1 March 1695; and all lieutenant-colonels are to be personally at their quarters under pain of being cashiered.

Doubtless there had been many casualties which needed filling up, though the fighting had been nil. It is clear that the success achieved by the officers of Cunningham's Dragoons in recruiting was not always equalled by other regiments. For the previous campaign, that of 1693, we read of a scarcity of recruits, and apparently the officers were slack. Captains were ordered to be cashiered if they did not complete their troops within a specified time, and an abatement of their pay proportionate to the deficiency in men was decreed after 1 May. With all that, it was however, forbidden most strictly to enlist Prest men for Flanders and the severest punishment is decreed to be meted out if the practise continues.

After the review by the King, the British dragoons, Cunningham's among them, went into cantonments in the villages between Ghent and Sas van Ghent, and here they remained till the opening of the campaign of 1695.

The most important, nay, even the only important, result of the campaign of 1694 was the capture of the town of Huy. In this Cunningham's Dragoons were not concerned; yet a brief account of it may be inserted, seeing that on its surrender the French were compelled to evacuate the whole territory belonging to the bishopric of Liege. Field-Marshal His Highness the Duke of Holstein Ploen commanded he allied army. The besieging force left the camp at Wouterghem (otherwise Aerschot) in detachments early in September, and established a blockade of the place by the 7th. Huy was a fortified town of which the defences had been much improved by the French during 1693-4. The allied army had a strength of about 20,000 foot and 3800 horse.

At the time of the investment the French governor of Namur was within the place, but under orders quitted it with a strong escort of dragoons, and after some difficulty reached Namur, leaving his baggage behind him.

The force defending Huy amounted to 1400 foot and a troop of dragoons under the command of Monsieur de Regnac.

At noon on 7 September the magistrates of Huy came into the allied camp and obtained terms by which the town was surrendered. This was by the leave of Monsieur de Regnac. The defending force then withdrew into the castle and forts. The town itself was occupied by four battalions of the allied troops.

Batteries were then erected by the besiegers. A condition of the surrender was that the allies. . . .

. . . .should not attack the castle by the town or faubourgs, nor cause any troops to pass the bridge in a body during the siege; nor any cannon, ammunitions or any other provisions of war; and that by virtue of this capitulation he should deliver the next day by six in the morning the keys of the town in the hands of the Magistrate.

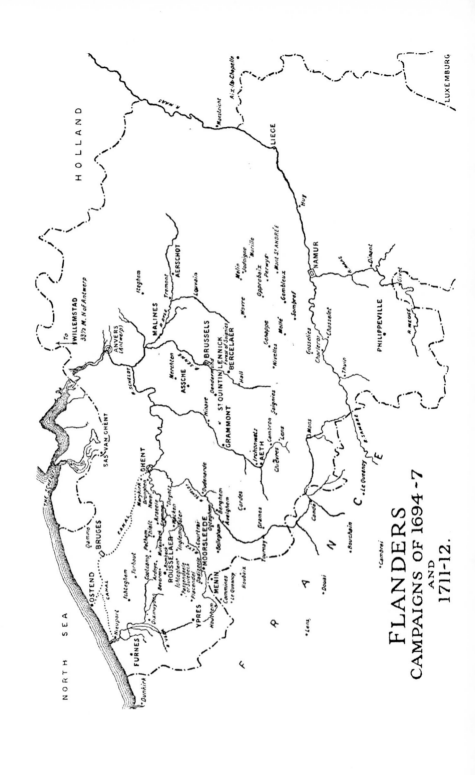

FLANDERS
CAMPAIGNS OF 1694-7
AND
1711-12.

A brisk cannonade on both sides began on the 7th, and continued with but few breaks until the 17th, when the garrison surrendered. The besieged sallied on the 7th, but were repulsed. On that day the besiegers finished bridge of communication at Taverne. On the 9th three guns were disabled in the Fort Picard and some miners killed. This was reported by deserters. Meanwhile the batteries of the allied army were being perfected and the heavy artillery placed in position.

Besides field artillery the siege train consisted of fifty-five battering pieces and twenty-eight mortars. These had been brought to Huy by water from Maestricht.

On the 10th the fire of the besieged slackened, as they were employed in covering themselves 'in their works with earth and hay, to defend themselves from our Bombs.' Next day the fire of the defenders was much hotter. The five batteries of the besiegers were completed.

On 12th September the fire on both sides was very hot, and trenches were opened at the Brandenburgh quarter against Fort Picard. Fog prevented bombardment for some time on the 13th. When it lifted, the fire of the allies was hotter than ever. Volleys directed against the bridge by the French caused a message to be sent that if this continued no quarter would be given, it being a breach of the capitulation. The well of the besieged was this day ruined, and the store of casks of beer and water destroyed.

On 14 September the fire of the besieged ceased. Fort Picard, in which a breach had been made, was stormed and carried. Fort Rouge next fell, as did the Tower of St. Leonard. Another tower surrendered at discretion. Of all these, however, the guns were spiked by the commandant, who escaped with about seventy men. A mine was sprung after his retreat, but did no harm. Many prisoners were taken. The loss of the allies was not more than ten men.

On the 15th the castle was summoned to surrender, and there was a cessation of hostilities for half an hour. The French, however, determined to hold out. During the following day the guns of the besiegers pounded the defences, and this was continued till 1 p.m. on the 17th, when the enemy beat the Chamade and asked to capitulate.

On the 18th the defenders were permitted to march out through the breach between nine and ten o'clock in the morning, and had much difficulty in getting over the ruins. The breach, in fact, was by no means practicable, and had it been stormed could hardly have been taken.

This marching out by the breach arose from the fact that in the Articles of Capitulation to 'march out by the breach' had been specifically named. When the besieged found how difficult it was to make their way over the ruins they stopped, and asked to be allowed to march out through one of the gates. Leave was however refused, and they had to scramble out as best they might. Terms of capitulation, of which many are extant, often form curious reading, but cannot, of course, be quoted here.

CHAPTER 2

Campaigns
1695-1697

We left the regiment cantoned for the winter at some villages between Ghent and Sas van Ghent. Mainly owing to the absence of King William in England the opening of the campaign of 1695 was delayed till the middle of May.

The King landed in Holland on the 14th of that month, and remained at the Hague until the 20th, when he went to Loo. Meanwhile, the allied armies began to take the field. One was commanded by the Elector of Bavaria and the Duke of Holstein Ploen, the other by King William in person and Charles Henry of Lorraine, Prince of Vaudemont. King William designed to operate in Flanders. When Cunningham's Dragoons left their winter quarters they marched to Dixmuyden in West Flanders, thence to Arseele. The King, who had left Loo for Breda, now proceeded to Ghent, where he was accorded a most enthusiastic reception by the burghers. At Ghent he remained but a few hours, and arrived at camp at Arseele about midnight the same day. On 31 May, in the morning, the King reviewed the English cavalry which was encamped upon the right wing, having arrived on the 29th.

Cunningham's Dragoons, according to d'Auvergne, were brigaded with Dopf's Dragoons and under the command of Brigadier Wynne, being posted in the 2nd line of the left wing. They were reviewed in the afternoon of the 31st, and are stated to have been, like the cavalry in general, in 'an extraordinary good condition.'

On 2nd June the army marched in four columns from Arseele towards Rousselaer dividing from the centre. Cunningham's

Dragoons proceeded by the high road. At Rousselaer the force camped. The army on the morrow marched to Becelaer, and for the first time came in touch with the enemy. A detachment of 400 dragoons, which had been sent out on the extreme left of the left flank, fell in with a party of the dragoons of the enemy. These they incontinently charged, defeated, and chased to the very stockades of Menin, returning to the army with twenty-three prisoners. Captain Stanhope, a volunteer, is recorded to have had his horse shot, but it is not clear to what regiment these dragoons belonged. From Becelaer, on the evening of his arrival, William rode out to reconnoitre the enemy's line, which was some three miles distant from the allied camp. He found the enemy's line 'very well palisaded, and the parapet very thick and strong, flanked with good redoubts, and cannon mounted for the defence of them.'

William returned to camp late that night, and found that not until midnight had the artillery, baggage, and rear-guard arrived. Marshal Villeroy, who had already sent forward reinforcements to the French army, was himself approaching rapidly with the main bulk of his force.

Prior to marching on the morrow William again rode out to reconnoitre. He had with him a considerable detachment. It was in fact a reconnaissance in force. The enemy, fearing an attack and assuming that the King's detachment was the vanguard of the attacking force, beat to arms. William then returned to the camp at Becelaer and countermanded the order to march which had been issued on the previous night. William's design was to compel the enemy to bring as many troops forward to defend the fortified camp as possible, though he well knew that it was next to impossible for any attack thereon to succeed.

Manoeuvring of a somewhat intricate nature now continued for several days. The Elector passed from Nivove to Houthem, towards the Scheldt. Next day he passed the river below Oudenarde. On the 6th he encamped near the enemy's line with his right towards Courtrai, his left near Veermande, and his headquarters at Castres.

Cunningham's Dragoons were now to have an opportunity of coming to hand grips with the enemy.

At Becelaer forage for the horses was by no means plentiful, and the dragoons were fully occupied in obtaining whatever sup-

CORNET JOHN BLAND
(LATER COLONEL)

plies they could. It was by this method that the forage was gathered when near the lines of the enemy. The dragoons dismounted and accompanied by infantry would go out and cut forage. Then, leaving the infantry to guard it, they returned to camp, brought out their horses and loaded up. By this precaution they foraged almost up to the gates of Ypres without losing a horse.

On 14 June King William received news that the French had conceived the design of intercepting the bread wagons which were coming laden from Bruges to the camp with the provisions for the army. The spot where they intended to fall upon them was Rousselaer. Accordingly, Lord Essex was ordered, with 500 dragoons, to join the convoy at Rousselaer, and Lord Portland, with 500 more dragoons commanded under him by Brigadier Wynne and sustained by some horse, was told off to intercept the enemy. Lord Portland received information that the French force designed to pass the night at the village of Moorsleede, and thither he made his way with the troops under his command. The enemy was found already established in the village, the streets of which they had managed to barricade with overturned carts and wagons. The strength of the French force, which was under the command of a lieutenant-colonel, amounted to 400 men.

Dismounting, the British dragoons at once attacked the barricades and speedily captured them. Several of the enemy fell in the assault; a captain and thirty men were taken prisoners, while the remainder bolted. With the British the Count of Soissons, a brother of Prince Eugene, was serving as a volunteer and bore good testimony to the briskness of the attack. The affair lasted about half an hour. On the British side a Lieutenant Webb, the brother of Lieutenant-Colonel Webb of the Guards, was killed. Brigadier Wynne was shot in the knee, and Captains Collins and Holgate were both wounded. The wound of Brigadier Wynne, at first not deemed serious, afterwards proved mortal, and that gallant officer died a few days later at Ghent. Besides these the British lost some men in killed and wounded. None of the three officers whose names have been given belonged to Cunningham's Dragoons. On the morrow the convoy of bread wagons arrived safely in the camp. almost simultaneously a force of some fifty of the enemy had been captured near Bruges.

This was, however, somewhat balanced by a success of the enemy on the night of the 15th, when the French fell upon an outpost near Sonnebeck and made several prisoners.

On the 17th the artillery of the allies proceeded from Becelaer to Rousselaer, marching by night. A few bodies of French troops were seen, but the escort prevented any attack from being made on the guns.

Next day the whole army marched to Rousselaer in two lines, King William remaining with the rear-guard. Marshal Villeroy meditated an attack on the rear-guard, but did not carry it into execution, and the allies reached their destination on 19 June.

Early that morning William left for the Meuse with a strong escort. The siege of Namur was now about to be undertaken. Namur was a strong place and was deemed by its garrison almost impregnable. On one of the gates was inscribed 'Reddi quidem, sed vinci non potest,' as much as to say they could restore the town to its owners, but that it could not be taken. The siege began on 3rd July; by 6th August Count Guiscard the commandant had surrendered. Meanwhile, Marshal Villeroy had suddenly fallen upon Dixmuyden and Deinse and compelled them to capitulate. Later, the officer who had been in command at Dixmuyden was tried by court-martial and decapitated by the common executioner. He was by nationality a Dane, and was accused of misbehaviour in the fact that he did not show fight. A British dragoon officer, by the way, was the only man who opposed the capitulation, and was thereafter promoted for that reason.

Cunningham's Dragoons, however, had not been engaged at Namur, but had formed part of the army under Charles Henry of Lorraine, Prince of Vaudemont, to whom was entrusted the duty of covering the besieging force.

The Prince of Vaudemont's army numbered about 36,000 men. His camp was at Wouterghem originally, but on 3 and 4 July he slightly shifted his ground to a position more favourable for defence. His right extended as far as the rising ground at Arseele, and his left towards Wacken and the river Lys. The whole line was entrenched. Here he awaited Villeroy, though the latter had more than double his force. The French army approached unmolested save by a Dutch major, who, posted with two hundred foot at Inghelmonster on the river Mandel, had entrenched and palisaded a neighbouring house which also possessed a moat. This plucky officer interrupted the march of the whole French army for some time, causing them considerable annoyance, and it was not until cannon was brought against him that he surrendered. Some of his captors wished to make an example of him for his temerity, but Villeroy, to his credit, 'approved of his courage and bravery, and ; pleased to applaud it.'

On came the French, and were drawn up in line of battle facing the Prince of Vaudemont quite early in the day. Nor was this all, for a strong detachment under Montal had got round on the right rear of the allies and had outflanked them. Villeroy might have attacked at once, and could have done so with every prospect of immediate and complete success, but, whether from over-caution or from a desire to make his victory absolute and complete he delayed. The Prince of Vaudemont, however, fully comprehended the dangerous situation in which he was placed. He therefore put into execution a clever manoeuvre by which he succeeded in withdrawing his army in a masterly manner, and practically right under the nose of his foe. That he was justified is amply proved from the fact that, had he remained to fight, it would simply have meant the loss of his entire army.

When Villeroy decided not to attack at once, he gave orders to Montal to take the posts of the allies which were existing on the rear of their right, between Arseele and Wincke. The Prince of Vaudemont became acquainted with these orders, and what he did to escape from the net by which he was threatened to be surrounded d'Auvergne tells us in detail. Let us quote his account.

> Our Army was then posted in the retrenchment, expecting the enemy; and though Montal had already pass'd Thielt, and was drawing near to Caulghen, Prince Vaudemont chang'd resolution, and thought it very hazardous to venture a battle which promis'd the total ruin of his army, and then, though the time was urgent and pressing, he immediately, with a most admirable judgement, resolv'd upon, and contriv'd a retreat. The Prince had very wisely provided for such an accident in the morning, by ordering all the baggage to load immediately and to march by Deinse to Ghent that it might not embarrass the motions of the Army. The Prince order'd first the cannon to be drawn off the batteries, and to march towards Deinse; which was done so secretly, that the enemies did not perceive it. He had wisely ordered the artillery to

be moving from battery to battery all the afternoon, so that when it went clear off, the enemies thought it had been but the ordinary motion. After, the two lines of foot march'd upon the left, along the retrenchment. To cover this march of the foot the Prince order'd a body of horse to come and post in the retrenchment; as 'twas quitted by the foot. The foot march'd with their pikes and colours trailing to conceal their march; neither did the enemies perceive this motion till the Cavalry mounted again, and abandon'd the retrenchment, and then the infantry was already got in the bottom between Arseele and Wouterghem, marching towards Deinse. At the same time that the foot were filing off from the retrench-ment, the Prince order'd Monsieur d'Auverquerque, with the right wing of horse, interlin'd with Collier's Brigade of foot to make a line facing Caneghem, extending himself from the Windmill of Arseele towards Wincke. This motion was to make Montal believe that this line was design'd to oppose his attempt upon the rear of our right; but his secret orders were to march off by Wincke to Nevel, and so to Gh-ent. At the same time that the foot march'd by Wouterghem and Deinse, my Lord Rochford who was posted with the left wing of horse and two battalions towards the Lys, made the rear-guard towards the left, with a line of foot on one side, and three squadrons of Eppinger upon the other. All this was contriv'd by the Prince, from the right to the left, that the Army disappear'd all at once, just as if it had vanish'd out of the enemy's sight. The Prince, and the Duke of Wirtemberg, and other Generals, kept to the retrenchments till all was march'd off; forming with themselves, domestics and attend-ants, a little body of horse, still to impose upon the enemy, and followed the army as soon as 'twas all got off. The enemies finding themselves cheated, did what they could to overtake and fall upon our rear: Montal particularly endeavour'd to fall upon that body commanded by Monsieur d'Auverquerque, which march'd off by Wincke to Nevel. He overtook the rear with some squadrons of horse and dragoons: But our defiles were good, and Brigadier Collier had order'd all the grenadiers of his brigade to the rear of all, to face the enemy from time to time as they advanc'd in their defiles; which was so well contrived that the grenadiers with their fire kept the enemies at a distance, and made the retreat good, and Montal could not do us the lest harm. When they had fail'd here they endeavour'd to fall upon the rear of our body of foot, which was brought up by Count de Noyelles Lieutenant-General. The (he) order'd a line of foot to advance, with some horse and dragoons; but the foot were already got so far that they could not hurt them. However two squadrons of their dragoons put green boughs in their hats,[1] which is our sign of battle, and spoke some French and some English, as if they had been some of our rear-guard: It was then the dusk of the evening; and with this stratagem they were suffered to come up close to our rear of foot, and march'd with them a little way, till they came to a convenient place, that they fir'd upon our rear, and then fell in with their swords. This put the first battalion in great disorder; but the other immediately facing about, oblig'd the enemies to retire. They killed us several men, and made some prisoners: The Lunenburgh Regiment of Luck suffer'd most on this occasion. And this is the only loss we receiv'd from the enemy in this great and renown'd retreat; which is as fine a piece of the art of war as can be read of in history, and which can hardly be parallel'd in it; which has shew'd more the art, conduct, and prudence of a General than if the Prince had gain'd a considerable victory: and this is the sense his Majesty was pleas'd to express of it in a letter he writ to Prince Vaudemont on this occasion.

Thus it was, then, that the Prince of Vaudemont saved his army and made good his retreat to Ghent. Had any disaster occurred to his force, it would have meant the raising of the siege of Namur.

On arrival at Ghent, Cunningham's Dragoons were detached with Rosse's Dragoons and twelve battalions of infantry under Lieutenant-General Sir Henry Bellasis to cover Nieuport, a seaport town, being between Bruges and that place. Here they remained until Marshal Villeroy advanced towards Namur in an endeavour to raise the siege. Cunningham's Dragoons then were moved to Brussels.

---

1. The troops of William III at the battle of the Boyne all mounted green boughs in their hats.

38

Meanwhile, Dixmuyden and Deinse had fallen, as has been mentioned, Namur had capitulated, and Brussels had been bombarded by the French, in the presence, of the allied army, who were unable to prevent this needless and wanton destruction of property. Brussels suffered much from the fire caused by this bombardment, the Stadthuys and the marketplace and several streets near being destroyed. The wealthy and trading part of the town, too, was most affected. Besides this a large number of churches were consumed, and their valuable contents, such as pictures, jewels, and plate, perished in the flames. As a matter of fact, this bombardment was a most wanton affair, causing needless suffering, and for no military reason whatever, except an attempt to draw off the forces then engaged in laying siege to Namur.

Later, Cunningham's Dragoons encamped for a short time on the Bruges canal, and then in October went into winter quarters in villages upon the canal of Ostend, in the Pays du Nord. The campaign of 1696 was decidedly uneventful. A plot had been formed for the restoration of James II, and to assist in its execution a number of French troops were marched to the coast, it being intended to transport them to England to assist the Jacobite rising. In consequence, twenty battalions of Dutch and British troops were sent over from the Continent in preparation for the expected French invasion. William was detained in England until the middle of May. The plot failed, but the measures adopted by the British Government to oppose the invasion crippled the army of the allies in Flanders, and practically prevented almost any operations of importance from taking place there. Still, as early as March, the Duke of Holstein, who commanded the allies in the neighbourhood of Namur, detached a force under Lord Athlone and another under General Cohorn, and sent them up the Meuse. Dinant and Givêt were closely invested, and the extensive French magazines at the latter place were totally destroyed. The detached force, having accomplished its mission, returned scatheless to Namur. Meanwhile, the bulk of the French army, which, under Villeroy, was encamped on the plains of Cambron, did nothing, and the remainder, then encamped at Rousselaer under Marshal Boufflers, was equally inactive. On the arrival of William he encamped with the Elector at Halle in order to cover Brussels, while the Prince of Vaudemont lay along the canal between Ghent and Bruges to oppose Boufflers.

Both armies faced each other during the whole campaign of this year and did nothing. In October both armies retired to winter quarters, and rumours of peace were rife.

The events referring particularly to Cunningham's Dragoons during this year are few.

On 13th May clothing arrived from England, being brought by water from Bruges to Marykirk. By this time it must have been badly needed.

King William had reached the Hague on 7th May, where he remained until the 13th, proceeding thence to Loo, and finally arriving at Breda on the 25th. He marched into the camp near Ghent on the 27th. In the afternoon of the 30th King William reviewed the English cavalry, and it is reported that Cunningham's Dragoons 'made a very good appearance, both man and horse being in very good order.' On 31 May many officers were promoted, among them being Colonel Cunningham, who was 'made Brigadier of Dragoons in the place of Brigadier Wynne, that had died the last year of his wounds received at Moorsleede.' In June King William was at Wavre, leaving that place on the 9th.

Cunningham's Dragoons were then serving under their old general the Prince of Vaudemont, being brigaded with the regiments of Eppinger and Miremont, the regimental command being still retained by Cunningham. They were not engaged in the attempt against the French cavalry on 19 June, an attempt which proved ineffectual. The regiment remained in its quarters near Ghent until 27 June, when it marched out and encamped along the Scheldt near to the citadel.

The allied armies during July were reinforced by a body of about 15,000 men and twenty-eight guns under the Landgrave of Hesse. This force had been marched thither through Germany.

During this month the Prince of Vaudemont, with a strong escort of 500 dragoons and 700 foot, journeyed in his coach with the utmost secrecy to the King's camp and arrived there on the 21st. His escort on the way was reinforced from the King's army. He returned to his camp on the 22nd by a more circuitous route, the shorter and more direct way he had taken on his outward journey not being considered safe.

Expectation was rife in the camp that some operation was at

length about to be actively undertaken, and that the army was about to decamp. It did decamp, but only to take up a new ground and on a more methodical plan, the reason being that the corn had now been garnered, rather early it is true, but this was by the special order of the Prince of Vaudemont. Hence it was possible to have the camp lines drawn upon a more regular system and more in consonance with the military ideas of that day. The Prince also insisted on all corn being threshed out and conveyed to his side of the canal.

The army was now cantoned along the canal of Bruges. Cunningham's Dragoons were again in the 2nd line of the left wing, and appear to have been encamped between the Halfway House and St. George, places it does not seem possible now to identify. At least, one account states that the Dragoons were so placed, while a few lines later it is stated that 'the English Cavalry lay between Marykirk and Ghent, upon the Left.'

During the month of August there were several small skirmishes, but nothing of importance occurred. Foraging parties were frequently sent out on both sides, and sometimes came into collision.

Towards the end of the month the camp of the Dragoons of Eppinger, Rosse, and Miremont, with whom Cunningham's were brigaded, was shifted to a spot between the right of the foot and Bruges. On 31 August orders were received for the dragoons to pass the canal and to encamp within a new entrenchment that had been recently constructed and finished on the previous night. This entrenchment extended 'from the Château of Coudekenken upon the right of St. Andreas his Cloister to St. Michael upon the left, which defended all the plain, this being the most open passage to Bruges.'

Villeroy had been very active in making a reconnaissance in force, and was expected to attack on this line, hence the entrenchment.

On arrival within the entrenchment the dragoons were encamped near the canal of Ostend on the right of the first line. On the same day Villeroy's army marched towards Torhout and the canal of Nieuport. Their right was at Torhout, Ichteghem towards the canals of Nieuport and Dixmuyden was on the left, and the wood of Wynendale in the rear of the centre of the camp. At Wynendale Villeroy took up his quarters.

On 1st September guns were placed in position within the entrenchment. Plassendael was now fortified, and four guns were mounted there. To aid in the defence of the entrenchment and also to annoy the enemy, the digue of the canal of Nieuport was broken down in two or three places, and the French got the benefit of a heavy flood of water. Early in September the French concluded a separate peace with the Duke of Savoy. From this peace the enemy expected to derive much advantage, hoping to extend its influence so as to include all Italy.

The strenuous efforts of the allies to safeguard the town of Bruges did not pass unrecognised by the magistrates of that place. They arrived one day in a body at the camp in order to compliment the Prince of Vaudemont on the success of his endeavours. Nor were the soldiers forgotten, for the worthy burghers brought with them a present of 400 barrels of beer! This was doubtless a most welcome gift, and duly appreciated by the men. The end of this uneventful campaign was now at hand. On 28 September the Prince of Vaudemont reviewed the English cavalry and dragoons at Moorbrugghe. The King had already left the army and was on his way to England. He sailed from the Hague on 4 October. On the same day the army of the Prince of Vaudemont began to file off towards its allotted winter quarters, the dragoons, Cunningham's amongst them, being located behind the canals of Ostend ad Bruges.

The Regiment was now about to lose its first colonel. On 1 October 1696 Brigadier-General Richard Cunningham was succeeded in the colonelcy by William, Lord Jedburgh, the eldest son of the Marquess of Lothian. In consequence of this change the designation of the Regiment became Jedburgh's Dragoons.

The events of this campaign are of comparatively little interest. In the month of May the opposing armies began operations. To the French a considerable accession of strength had come owing to the fact that the peace with the Duke of Savoy had released a large army.

This army, under the command of Marshal Catinat, then joined the forces already in Flanders under the command of Villeroy and Boufflers. The French host drew together and encamped on the plains of Cambron, and totalled 133 battalions of foot and 350 squadrons of cavalry.

King William encamped at Bois-Signior-Isace, his army being about half the size of that of his opponents, and by means of spies kept a most careful watch on the movements of the French.

It was a most critical juncture, for negotiations for peace were in progress at Ryswick near the Hague, the French plenipotentiaries having already arrived there. William knew that if the French could steal an advantage by any unforeseen move it would most adversely affect the terms proposed for the peace. Hence he was more than ordinarily watchful, as at Ryswick procrastination was the order of the day and its intention was pretty evident. Bois-Signior-Isace was about five leagues from the French lines.

Suddenly, early in May, the French army moved forward towards the King's camp. He had however, obtained previous information of the design, and being unable to fight with any hope of success, he retreated towards Promel, where he was about to encamp and take up a position. But the French advance was a blind, and they suddenly deflected their march in the direction of Brussels.

Now, if Brussels fell into the hands of the French it would have been a most serious matter, besides the question of pillage. With all speed, therefore, William set his army in motion, and a race for that city began.

Marching night and day King William outstripped the enemy, and having passed through the wood of Soignies, arrived at Brussels some hours before the French. A strong position was taken up at Anderlecht, where a camp was formed. A strong entrenchment with several redoubts was hurriedly constructed, and by this means both the army and the town were secured.

The French, finding themselves foiled, stopped short at Halle, and a force was detached under Catinat to lay siege to Aeth, though it was confidently expected that peace would be concluded before the town fell This expectation was realised; still the town was taken.

Aeth had been fortified by Vauban, and it is curious to note that that distinguished master of the art was himself engaged in directing the siege.

Peace was signed between England, Spain, Holland, and France on 20 September and between France and the Empire on 30 October.

Both armies immediately quitted the field, thankful, no doubt, that this protracted war was at last at an end. The British troops were quartered in Ghent and Bruges till transports could arrive on the coast to take them to England or Ireland, Ostend being the chief port of embarkation.

A passage in a book published in 1748, which purports to be the diary of a Captain Robert Parker, who served through several campaigns both in Ireland and the Low Countries, is so curious as to be worth quoting. It is as follows:

> It is worth observing how great a number of Troops were quartered in the single town of Ghent at this time, and the affair was so happily conducted, that the burthen on the inhabitants was very inconsiderable. This was owing to the contrivance of a poor button-maker who made a fortune by it.

Unfortunately, he does not tell us what the contrivance of the poor button-maker was. As a matter of fact 29,650 men were collected together in Ghent, of which not half were British. Here again we find a discrepancy between Cannon and the Regimental Record. Cannon says:

> The Regiment, bearing the title of Jedburgh's Dragoons, served the campaign of 1697 in Flanders with the army commanded by the Elector of Bavaria, and was formed in Brigade with the Regiments of Nassau-Sarbruck, and Opdam, under the orders of Brigadier General Pyper. It took part in several operations; and in May joined King William's Army in Brabant, but subsequently returned to Flanders.

From the Regimental Record we find that in April Jedburgh's Dragoons were stationed at 'Merchtein and Asche near Brussels with the Elector of Bavaria,' and that on 17 May they were 'at the Camp of St. Quintin Linneck with King William's Army.'
The next entry reads oddly, and is as follows:

> The Regiment was not at any time engaged with the enemy during its stay in Flanders!

I transcribe this last entry to draw attention to some of the difficulties likely to occur to those occupied in writing the history of a regiment.

CHAPTER 3

Home Service & Campaigns
1697-1714

The Regiment accordingly returned home, and from Marching Orders W. O. 5, No. 10, we learn:

William and Mary
Jedborough
Our W and P is that you cause our regiment of dragoons under your command now in the river to land forthwith at Deptford and Greenwich and march the same day to our borough of Southwark from whence they are to continue their march according to the Routes they shall receive in that behalf to Berwick, and from thence into Scotland pursuant to such orders as shall be given them by our Council of that Our Kingdom or the Commander in Chief of our forces there. And the officers &c. &c.
Given &c 9 Feb. 1697-8
By Order
W. Blaythwait
To Ye Lord Jedborough.
Ye routes entered six leaves further.

The route was as follows: the Regiment started on Wednesday 16 Feb. 1697-8 and arrived on Tuesday 19 March at Berwick. For the march the Regiment was divided into four divisions of two troops each. All the divisions took the same road: Waltham Abbey, Waltham Cross, Cheshunt, Ware, Hoddesdon, Royston, Huntingdon (one day's rest), Peterborough, Stamford, Grantham (one day's rest), Newark, Tuxford (one troop), Worksop (one troop)—evidently half the Regiment went to Worksop and half to Tuxford—Doncaster, Pontefract, Newbridge, York (one day's rest), Boroughbridge (one troop), Easingwold (one troop), Northallerton (one day's rest), Darlington, Durham, Newcastle (one day's rest), Morpeth, Alnwick (one day's rest), Belford, Berwick. This is the route given for one division, and it is given in precisely the same order for all four.

This corrects the account given in the MS. Regimental Record which states that the Regiment landed 'at Harwich in December.' Cannon states the same thing and adds that they 'proceeded to London, where they occupied quarters for several weeks; at the same time their numbers were reduced to a peace establishment.'

It is possible that the numbers were reduced, but there is no documentary evidence to prove the fact; consequently, we can but mention the matter without offering any suggestion as to the amount of the reduction if any. The expenses of the march to Berwick are given in the Regimental Record, and are as follows:

For 4 wagons for the Regiment in four divisions 320 miles at 8d per mile	£43 13 4
For fire and candle in their several quarters in their march from 9 Feb. to 20 March being 39 days at 2/- a day for each division	£15 12 0
	£183 9 0
Poundage	£8 4 4
	£191 13 4
Deduct overcharge for fire and candles	£21 19 8
Total	£169 13 8

The last entry, 'Deduct overcharge for fire and candles,' cannot well be, explained, seeing that the amount deducted, £21 19s. 8d., is greater than the whole charge, £15 12s.

The Regiment evidently brought its horses back from Flanders, otherwise a far greater length of time would have been consumed on the march between London and Berwick.

From this date, i.e. 19 March 1797-8, until 15 March 1707-8, it appears impossible to discover the whereabouts of the Regiment, beyond the fact that it was in Scotland. Nor, indeed, is there any record as to its proceedings. All possible manuscript records have been diligently searched in London, but without success. The fact is, that there is a great dearth of information as to regiments at this period when on the Scots Establishment. The task of searching at random collections of contemporary letters for possible information would occupy a lifetime, and in all probability with but little commensurate result. One might find references to the Regiment by its colonel's name, but more often cavalry would simply be called dragoons, without any clue being given as to what particular regiment of dragoons the allusion referred.

The author has been already some considerable time engaged in an endeavour to fill up this gap, but unfortunately without result. He has therefore reluctantly been obliged to abandon further systematic quest. From McPherson's *Secret History of England*, edition of 1775, vol. 2. p. 7, Mr. Scott's relation, 'An Account of the State of Scotland', in July 1706, we get the following:

> The Earl of Lothian's regiment of dragoons (as I remember) consists of six companys, each company, including serjeants, corporals, and drummers, is thirty men. The colonel's character is already given. The lieut.-colonel is son to Polwarth, now called Earl of Marchmont. When the late Earl of Hume listed, this lieut.-colonel was thought well-affected, and very much under the influence of Hume; but what to say of him now I know not. The major of the regiment, John Johnston of Westraw, is reported to have loyal inclinations, being much managed by his very loyal lady, whom few of any side must trust.

Upon this extract the following remarks may be made.

William Kerr, second Marquess of Lothian, succeeded to the title in 1703, and never commanded the Regiment as Earl of Lothian. Of course companys should be troops. For the character of the colonel see Appendix 1

The Hon. Patrick Hume, afterwards Lord Polwarth, was Brevet-Colonel 30 March 1704, Colonel 28 April 1707, and died in 1709. What the late Earl of Hume had to do with the matter is not apparent. The father of Lord Polwarth had, it is true, been suspected of Jacobite leanings and wrongfully, but Colonel Lord Polwarth was always an anti-Jacobite (see Appendix 1).

From this extract it would appear that the strength of the Regiment, at that period was 180 men.

Also we see that it had been reduced from eight troops to six. In only one instance can the quarters of the Regiment be fixed during the period between March 1697-8 and March 1707-8 ; and that is for the month of September 1703, when the six troops of the Regiment were inspected as follows:

Johnston at Dunse, 10 September.
Guidett at Queensferry, 22 September.
Douglas at Costorphine, 28 September.
Drummond at Jedburgh, 11 September.
Preston at Inverkeithing, 29 September.
Polwarth at Greenlaw, 10 September.

Dated from Whitehall, 15 March 1707-8, is a letter to Lord Polwarth and others to 'increase his Regiment by 129 Private Men.' This, too, is curious, for Lord Polwarth was not gazetted to the colonelcy of the regiment until 28 April 1708.

On the same date the troops were stationed at Kelso, Dunse, Coldstream, Queensferry, Pittenweem, and Greenlaw. During the month of may they shifted to Queensferry, Costorphine, Greenlaw, Kelso, Dunse, and Jedburgh.

From April 1708 the establishment of the Regiment was increased from thirty-two a troop to fifty-four; at least so we learn from the Regimental Record. and it is probably an incorrect version of the letter of 15 March 1707-8 quoted above. The augmentation was caused by the threatened French Invasion, an invasion designed to assist the Jacobite cause. It failed.

In another letter to Lord Polwarth, also addressed from Whitehall and dated 16 July 1708. he is directed to transmit 'an exact list of Officers in his Regiment and the date of their Commissions.' From an 'Abstract of the Quarters where the several troops of Dragoons in North Britain are Quartered,' we read of The Lord Polwarth's Regiment as follows:

April 2nd 1709
Polwarth
The Colonel's troop at Musslebourgh.
The L.-Colonel's. troop at Kelso.
The Major's troop ordered from Falkirk to Coldstream.
Captain Prestoun's troop at Jedborough.
Major Guydet's troop at Dalkeith.
My Lord Torpichen's troop at Easter and Wester Calders.
(W. O. 5, No. 15.)

The next paper is dated St. James's, 31 May 1709.

Routes
Route for one troop of the Lord Polwarth's Regiment, of
Dragoons from Kelso to Polwarts.
Ditto from Coldstream to Polwarts.

This would appear to refer to the troops of the Lieutenant-
Colonel and the Major. Apparently another troop went to Polwarts, as we find from an almost contemporary paper.

Route. Polwarths.
Polwarts....................3 Berwick.
Easdell......................1 County of Easdell.
Gortoun....................1 (no destination given).
Cather and Binney...........1 Midlothian.

This appears rather differently in the MS. Regimental Record,
dated 31 May 1709.

1 Troop from Jedburgh to Easdell.
1 Troop Dalkeith to Polwarth.
1 Troop Kelso to Polwarth.
1 Troop East and West Calder to Gortoun.
1 Troop Coldstream to Polwarth.
1 Troop Musselburgh to Cather and Binney.

On 18 October 1709 a troop was ordered to be moved to 'Cupar
in Angus,' but from what place or which troop of the Regiment is
not stated. Three other troops are ordered to Kelso, Dumfries, and
Crieff (W. O. 5, No. 16).

In most of the marching orders and routes of those days you
find the following inserted:

To rest the Sundays and every third or fourth day on their
march as the Officer in Chief shall see cause.

A route given in the Regimental Record, dated 18 October,
accounts for the movement of four troops.

One from Newbattle to Cupar in Angus
One from Greenlaw to Kelso.
One from Greenlaw to Dumfries
One from Dalkeith to Crieff.

This is evidently the order given above (W. O. 5, No. 16).

Lord Polwarth had now retired owing to ill-health, and he was succeeded in the command by Colonel the Hon. William Ker, the younger son of the second Earl of Roxburghe and brother of the first Duke of Roxburghe. In documents his name frequently occurs as Kerr, but it is in these days more correctly Ker. Colonel the Hon. William Ker obtained the colonelcy of the Regiment 10 October 1709.

The Regiment, now known as Ker's Dragoons, was again about to proceed upon foreign active service. Early in January 1710-11 definite orders were received for the Regiment to proceed to Flanders. With regard to this subject many papers exist, and copies of them are here produced when necessary.

Whitehall
5th January 1710/11
Sir, It is Her Majesty's Pleasure that you do appoint an Officer forthwith to attend the Commissioners of Transportation to indent for the number of men and horses of your Regiment which are to be embark'd as well at Leith as at Harwich, the said Commissioners representing that they are not to able proceed in this service till the same is done.
I am, Sir, your most humble Servant,
G. Granville
To Col. Kerr

It is curious that this order should precede that which follows, which is evidently the official order to proceed on active service abroad.

Whitehall
1 February, 1710/11
Sir, Her Majesty having thought fit that the Regiment of Dragoons under your command do serve this next campaign in Flanders, I am commanded to acquaint you therewith, that the same may be in a readiness to march and embark as soon as transport ships can be taken up for them, And in order thereto I am to desire you to cause an Account to be forthwith return'd to me of the exact number of men and horses for which transports are to be provided, distinguishing the numbers to be embark'd in Scotland, and

how many men and horses you will have here that may be ship'd at Harwich. I am Sir your most humble Servant
G. *Granville*
Col. Kerr
The like letter to Sir Richard Temple Bart.

Whitehall
5th February, 1710/11
My Lord, Her Majesty having thought fit that the Regiments of Dragoons of Sir Richard Temple & Col William Kerr now in North Britain be embark'd for Her Service this next campaign in Flanders, I am to desire your Lordship to let me know at what distance the said Regiments do now lie from Leith, where it is design'd they shall be embark'd and what places nearer to Leith are proper to march them to, so as that they shall be the more in a readiness of going on board, when shipping shall arrive there to receive them. I am Your Lordship's most humble and obedient Servant,
G. *Granville*
Earl of Leven

W. O. 5, No. 17
Anne R.
19 Feb. 1710/11
Whereas we have thought fit that our Regiment of Dragoons under your command be forthwith embark'd for our service in the Low Countries—Our will and pleasure is, and in order therein you cause the several troops to march forthwith from their present quarters in such manner as shall be directed by Our Right Trusty & Right Well-beloved Cousin David Earl of Leven, Lieut. General and commanding our forces in North Britain, and be so disposed of as follows: viz two troops at Dalkeith, one at Kincross, one at Cupar in Fife, one at Dysert and one at Calder, or at such other places as the said Earl shall judge most convenient, from whence they are to march to Leith and embark on board such ships as shall be appointed to receive them upon notice from our Commissioners of Transportation or their agent that shipping is ready for that purpose. And in the meantime the Officer is to take

care that the Soldiers behave themselves civilly and duly pay their landlords. And all Magistrates &c &c.

Given at Our Court at St. James's.

B. H. M. C.

G. Granville

To our T. & W. William Kerr Esq. Colonel of one of our regiments of dragoons, or to the Officer in Chief commanding the Regiment.

The Regiment duly went into the quarters specified. There were however, both recruits and horses in England, and these were dispatched from London to Harwich. The route given is Burntwood (Brentwood), Witham, Manningtree, and Woodbridge, 'Hence to march and embark at Harwich.' Woodbridge is quite near to Harwich. The old local saying of this ancient but decayed place is 'Woodbridge was a town when Harwich was a village.' Harwich folk object to the remark!

The next entry refers to an advance of pay to the troops, and enjoins the strict payment of all debts before embarkation.

Whitehall

22nd February, 1710/11

Sir, The Lords Comms. of the Treasury having directed a months pay in advance to the troops that are ordered from Scotland, whereof your regiment is one; I am to signify to you Her Majesty's pleasure that all due care be taken by your Officers for discharging their quarters, and that they do not leave any debts upon their marching.

I am Sir, your most humble servant,

G. Granville

Col. Kerr

The next document refers to the clothing destined to be dispatched to the troops, and also for the conveyance of the horses &c. for Flanders, and gives the date of embarkation at Harwich.

Whitehall

5th March, 1711/12

Sir, My Lord D of Ormond having appointed the Commissioners of Transports to have the ships ready to receive the clothing of the Regiments in Flanders on board on Monday next in the river of Thames and order'd the Transports to fall down the river on Thursday foll., that they may be at Harwich on the 15th inst: which day his Grace has appointed for the embarkation of the riding horses &c. for Flanders, I am commanded to give you notice hereof that you may take care to put the clothing of your Regiment on board in the river & to have your recruit horses &c. ready to embark at Harwich on the 15th inst. for which you will receive the necessary routes in time.

I am Sir, your &c.

Lansdowne

Col. Kerr

Following upon this, the next day another order arrived which concerned a draft of thirty-six men and horses taken from the Earl of Hyndford's Regiment of Dragoons.

Whitehall

6th March, 1710/11

Sir, Her Majesty having order'd a draught of thirty-six men and horses, such as are fittest for the service to be made out of the Earl of Hyndefords Regiment of Dragoons and incorporated into the several troops of your Regiment for completing the same to sixty men in each troop, servants included. It is Her Majesty's pleasure that you appoint an officer forthwith to receive the said additional men and horses, which you are to cause to be embark'd with your Regiment, upon the arrival of the transports at Leith, and to be paid as belonging unto your Regiment from the day of their being deliver'd over to you.

I am Sir, yYour most humble Servant

G. *Granville*

Col. Kerr.

The order contained in the last paper was duly executed; and its execution is referred to both by Cannon and the Regimental Record. '1711. The troops were increased from 54 to 60 each from March 1711.' The increase of thirty-six men would, of course, provide six extra men per troop, as the Regiment then consisted of six troops.

During the month of March, the Regiment, which it will be remembered as then quartered in places near Leith for the purpose, embarked upon the transports at that port.

Ill luck in the shape of contrary winds, however, prevented the flotilla from sailing at first. When at length, towards the end of the month, the vessels put to sea it was found impossible to continue the voyage, as the wind was still unfavourable. In consequence the ships put back into the Firth of the Forth, where they were detained until the end of April, when at last they sailed for Holland.

For details as to the coarse of the campaigns in the Low Countries Ker's Dragoons were serving at home, the reader must be referred to military histories which treat of that particular period.

Suffice it to say that, under Marlborough, the allied armies had pressed back the enemy gradually until the theatre of war was approaching the borders of France. To meet this danger Louis XIV had assembled; an enormous force.

The advisers of Queen Anne in consequence increased the British troops, and among those who were dispatched abroad on active service Ker's Dragoons were numbered, as we have already stated.

Marlborough, who had been in England, embarked for Holland on 18 February, and arrived at the Hague on 4 March. His first move was to place detachments from all the garrisons 'along the Scarpe and between that river and the canal of Douay.' He also established vast magazines of provisions and other necessaries at Tournai. About the middle of April the French began to assemble near Cambrai and Arras. Marlborough arrived at Tournai on 26 April.

The Emperor Joseph died on 17 April, and this delayed the arrival of Prince Eugene, who did not join the Grand Army until 23 May, on which day he was present with Marlborough and the rest of the generals at a dinner given in commemoration of the anniversary of the battle of Ramilles. This is the earliest reference to a military anniversary dinner the writer has met with.

Prince Eugene did not, however, remain long with the army, as he had to return to attend to the interests of the Archduke Charles, who had been elected Emperor. A considerable corps of the imperial troops were also withdrawn for the same purpose. The great event of the campaign was the capture of Bouchain.

Bouchain was invested on the right bank of the Scheldt on 7

August; but the investment was not completed on the other side of the town until the 18th. Bouchain surrendered on 14 September, the French army, which numbered 100,000 men, being perforce spectators of the fall of the place from their position between the Scheldt and the Sauzet. They even had a corps across the Sauzet at Wavrechain, almost within musketry range of Fagel's entrenchments.

It was most mortifying, but in so masterly a manner had this siege been conducted by Marlborough that they could do nothing. It may be noted that this glorious achievement was the last act in the military life of the great Duke of Marlborough. For political reasons, or rather from that curse of England then and now, party reasons, Marlborough was removed from his command.

During the retirement into Holland at the conclusion of the campaign the young Prince of Orange was unfortunately drowned while crossing the ferry over the Hollands Diep at Moerdijk. From his posthumous son the reigning family is descended.

For Ker's Dragoons this campaign was absolutely uneventful. Detained, as has been related, by the weather, the Regiment did not arrive in Holland until the allied armies had taken the field. After being cooped up in the so-called transports, which were really only adapted trading vessels, and ill adapted at that, for the conveyance of cavalry, both horses and men were out of condition and needed a period of rest for recuperation. Ker's Dragoons were therefore ordered to remain in quarters in Holland for a brief space to recover after the voyage.

Orders were then received for them to march to Brabant en route for the theatre of war, which was then very near the French frontier. They proceeded on their march for some distance; when near Brussels an order arrived for them to halt and go into quarters in that city.

Here they remained in reserve till the end of the campaign of that year. Probably through losses by sickness during the year it became needful to reinforce the Regiment. An undated paper, in which, however, reference is made to 'Quarter Orders of 17 September 1711,' gives a 'Route for the riders and horses of Col. Kerr's Regiment of Dragoons from London to Sudbury.'

The route is Waltham. Hatfield, Bradfield, Sudbury. The paper is signed 'Lansdowne.' Another paper directs their embarkation at Harwich after marching from Sudbury.

For the campaign of 1712 Prince Eugene, with whom was associated the Duke of Ormond, held the command. The regiment formed part of the army which was under the command of the Duke of Ormond.

In the spring an advance was made towards the French frontier, and an invasion of Picardy was in preparation. The siege of Quesnoy was first undertaken, for which purpose General Fagel was again detached, having the command of thirty battalions and eighteen squadrons. The siege began on 7 June and the place surrendered on 4 July. Its garrison consisted of 2700 men.

Peace negotiations were now in progress. The Duke of Ormond withdrew from the allies, taking with him the British troops. On 17 July he marched to Ghent and Bruges. Troops were sent from England to garrison Dunkirk, which town had been placed in the hands of the British Government as a pledge of the sincerity of the French king. This separation from the allies was as unjust as it was cruel and impolitic, for it left several smaller German States at the mercy of their enemy, and be it remembered that they had, as it were, been forced into the war by British influence. Now they had to make the best terms for themselves that they could.

For the winter of 1712 Ker's Dragoons went as usual into quarters. The Treaty of Utrecht was signed in February 1713, and by the middle the summer Ker's Dragoons had received orders to return home.

There are a few papers extant which give information as to the Kerr's Dragoons from the Low Countries to home. In the first place the Regiment was marched to its port of embarkation, Dunkirk. Here orders were received for it to proceed to Ireland.

Meanwhile, in pursuance of the policy of disbandment which followed on the Peace of Utrecht, the British Government had determined that Ker's Dragoons should be broken. But why they should have been sent to Ireland for this purpose passes comprehension. The Regiment originally of Scottish nationality was even now largely so, and the remainder was of English blood. Why, then, turn all these men adrift in Ireland of all places? But it was done. The order is as follows:

Anne R.

Our will and pleasure is, that upon the embarkation of our regiment of dragoons under your command, for our kingdom of Ireland, you cause all the horses belonging to the non-commissioned officers and private dragoons thereof to be delivered over to such person or persons as shall be appointed to receive the same; your said regiment being to continue unmounted until further orders. Given at our Court at Kensington, this 6th day of June, 1713, in the 12th year of our reign.

By Her Majesty's Command,
William Wyndham
To our trusty and well-beloved Colonel William Kerr, commanding one of our regiments of dragoons in Flanders.

With regard to the reduction of the British troops in Flanders, from a paper dated 2 June 1713, we learn that 'all British troops were reduced in Flanders, and afterwards by Degrees all were dispersed from thence to England.' Only the Royal Regiment of Ireland, commanded by Colonel Moses Leathes, held Ghent till 25 February 1715-6, and then delivered it up to the German troops and returned to England.

The reason why the Regiment was dismounted at Dunkirk is this. After their adventures in Spain the Royal Dragoons were returning dismounted, and the Government purposed to mount them on the horses of Ker's Dragoons. The Regiment arrived at Dunkirk for embarkation to Ireland on 15 August. The horses were shipped for Dover, and the men had sailed for Ireland by 21 August. A number of the men, however, accompanied the horses to Dover to take charge of them until the arrival of the Royal Dragoons. The horses and the party of dismounted dragoons duly arrived at their destination.

There appears to have been some trouble with the dragoons who were sent to Dover. Dated July 1713, we read a letter from Wyndham to Sir James Abercromby. It appears that nobody seems to have known how the dragoons thus sent to Dover were to be subsisted. The men seem to have been left without either food or pay themselves or forage for their horses. Wyndham orders that:

. . . . the Dragoons who brought over Kerr's horses are to be subsisted till they arrive here,' and he adds, 'I believe it will make them easy.

This really looks as if there had been trouble.

In a letter dated 3 August 1713 from Whitehall, Wyndham asks Sir James Abercromby for an account:

. . . . as to the equivalent the Officers desire to have made for their servants having hitherto kept their Companies effective. You will please transmit me a Stated Account thereof together with that for subsisting the reduced Dragoons who take care of Kerr's Horses, and the same shall be layd before Her Majesty.

In both these papers there is a conflict of dates with the 21 August and 15 August of the MS. Regimental Record. If the dates of the State papers are correct, and they must be believed to be so, it is evident that the entry in the Regimental Record was carelessly made. Besides, the horses were already in England by 25 July and had been some days.

As a dismounted regiment Ker's Dragoons remained embodied until the spring of 1714; the precise date is, as has been already stated, unknown. In the spring of 1714 it was broken. At the same time the Royal Dragoons and the Scots Greys were augmented, and a number of men from Ker's Dragoons were either transferred to, or on discharge enlisted in, these two regiments.

Thus ends the first portion of the History of the 7th (Queen's Own) Hussars. In future pages will be told the services of the Regiment at home and abroad after being re-embodied on 3 February 1714-15. A few interesting details as to military administration at this date may be inserted here, we think, without impropriety, more particularly as some of them have not, we believe, been hitherto included in any regimental history.

The methods of obtaining recruits were various. The first was the free recruit, who, either from a desire to serve, or for private reasons, such as   trouble, domestic or amatory, or a wish to avoid arrest for some delinquency, of his own deliberate purpose joined the army. This class of recruit, however, was never obtained in sufficient numbers to fill the ranks. Another method was by beat of drum. In this case the local authorities for general army purposes, or the officers of a regiment for the ranks of their own particular corps, by hook or by crook obtained men to augment the forces of the Crown.

The regimental method was somewhat after this fashion. A crowd having been collected by the aid of that somewhat noisy instrument the drum, possibly assisted by a fife, a serjeant by verbal exhortations, during which the fairytale element as to the advantages of soldiering was eminently to be discerned, endeavoured to inflame the martial susceptibilities of his audience. His exertions doubtless produced a thirst, and in the consumption of fluids stronger than water at, for the moment, apparently his cost, success was often his reward. With aching heads on the morrow thousands of young fellows have in the course of years found themselves duly captured as food for powder.

This recruiting by beat of drum in later days, when bands were established, developed the system of recruiting by band, in which the sight of marching troops through villages and towns accompanied by the lively strains of the band was found to be most successful. As late as the year 1883 it is on record that the Montgomeryshire Militia employed this means of filling its ranks to the required strength—the recruits thus obtained being numerous and of good quality, and they took the Queen's shilling readily.

Yet there are people to be found who would desire to see bands abolished in the service. In France the reduction in bands has been tried. It was a failure; and to the re-establishment of bands in that military country and to the institution of military parades through the streets is largely due the marked revival of the ancient martial spirit in the French nation. But in early times in England there were other methods of raising men. Prisoners for debt (and the prisons were full of these unfortunates in those days) could purchase their freedom by enlistment— nay, could almost be compelled to do so, no man under the age of forty being exempt; over that age he could plead and obtain the benefit of an Act of Parliament. There is a paper dealing with this dated 1695-6. Men could also be obtained by being pressed, but pressed men were forbidden to be enlisted for Flanders. It was found that forced service did not produce a good class of fighting man. The enlistment of Catholics, in Ireland especially, was absolutely forbidden, and later on the enlistment of Irishmen themselves was ostensibly completely banned. Both these edicts were absolutely a   dead letter, one method of evasion being to enlist men in Ireland, to send them over to Scotland, and bring them back to Ireland dressed in Scottish garb as Scots. With regard to the non-enlistment of Irishmen and to show how this regulation was simply ignored, a case has been known in which out of the numbers composing a regiment (much reduced, indeed it totalled only one hundred and thirty-eight in strength) no fewer than one hundred and twenty seven were Irishmen, whose enlistment was contrary to law. It is however, curious to note that in the 7th (Queen's Own) Hussars in early times the enlistment of Irishmen in the ranks was very rare indeed. Among the commissioned and non-commissioned officers the same conditions obtained. Why this was so we cannot decide, unless it was owing to the fact that the 7th Hussars were never one of those regiments who passed long periods in Ireland upon the Irish Establishment. The fact, however, remains. A favourite method of defrauding the Government was this. It was illegal to enlist an apprentice. An apprentice, however, desired a holiday, and obtained it thus. He would enlist, spend the bounty money, and then in due course be claimed by his master. All the latter had to do was to produce his indentures; and the professed would-be soldier was delivered up to him. Very large sums of money were at one time annually lost to the Government by this ingenious fraud.

The meshes of the enlistment net were also used to entrap felons. In fact, convicted felons—and the punishment for felony then was death—could often escape the consequences of their misdeeds by enlistment. But in the conditions of service there was this difference, whereas those who enlisted freely could join the regiment of their choice, prisoners for debt and felons were sent to whatever regiment the Government desired to send them, and theirs was service for life, and a service for life usually in some hated and unhealthy foreign station, where in a bad season they died like flies.

Deserters from one regiment often re-enlisted in another, chose their own regiments, and were careful, of course, to avoid proximity to their old ones. If detected, remission and severe punishment awaited them. Another point worth notice is that pay was often terribly in arrear.

Some time after he had been appointed colonel of the regiment now known as the 7th (Queen's Own) Hussars, but then styled Cunningham's Dragoons, we find the petition of Colonel Richard Cunningham and the officers of his old regiment of foot relative to the arrears due to their late regiment. These amounted to no less than £17,000, and the officers volunteered to compromise the debt for an immediate payment of £5000. If this was the case with a man like Cunningham, who was in favour with the Government and possessed moreover of influential friends, how much more was the less favoured officer likely to suffer! Hundreds of other examples could be cited.

And here one cannot but record a meed of praise to the measures which were taken by either William III or his advisers to ameliorate the state of affairs at that time. If these were due to the King, the credit is his; if they were initiated by his officials, he was to be congratulated on the possession of most capable servants. Let us cite a few examples, all derived from the State Papers. Officers are directed to take care that soldiers 'behave themselves civilly and orderly' to landlords and others, and the officers are to be held responsible for breaches of decorum, and vice versa.

It must be remembered that in those days billeting was most unpopular, and with civilians soldiers as a body were held in the greatest dislike. That the aversion was to a certain extent mutual is not to be wondered at, but the reasons for this dislike need not be discussed here. A relic of this aversion existed not so many years ago in the use of that objectionable term a common soldier—that this form of speech is now practically obsolete is a matter for congratulation. A soldier we know, a private soldier or a private we know, but of a common soldier we now have no knowledge, and resent the application of this term to any man who has the honour to wear his Majesty's uniform. But, as has been related above, King William III look measures to bring about a better understanding. How that King attempted to regulate the pecuniary relations between soldier and civilian in those barrackless days we will briefly tell below.

The superior officer in each quarter is ordered to sign the accompts of quarters when 'ye Landlord requests him.'

No trust is to be given to any soldier in quarters, and to prevent this in the case of foot 3s. per week was to be paid in two equal payments, one at the beginning and the other in the middle the week, to each soldier. For guards and fusiliers the amount is 3s. 6d.. The payment to dragoons is not stated. The money was to be paid without any deduction whatever, and officers were to account every two months with each man for 6d. a week extra, allowed for providing necessaries for each soldier to which the 'Off reckonings or residue of their pay hath not formerly been liable'. These Off reckonings were to be employed by:

. . . . the Colonels of each respective regiment for the clothing and poundage and of other remaining expenses for which due account is to be made for every soldier at each clothing.

This was a step in the right direction, and was meant to counteract the evil effects of the Proprietary Colonel system. Unfortunately, in later times it failed to do so. To prevent riots and the illicit use of weapons both officers and soldiers were forbidden to wear a dagger or bayonet at any other time than when such officers or soldiers were on duty or under arms. This order was to read at the head of each respective regiment, troop, or company. It, of course, did not affect the wearing of swords by officers, but dirks and bayonets (the latter at the time a part of the dragoon equipment) were banned. It is interesting to read a paper which details of the feeding of men and horses on board ship, and this scale of food was probably that in vogue when Cunningham's Dragoons sailed for Flanders.

For each man *per diem* one pound of bread, half a pound of cheese, and two quarts of beer or one quarter of a pint of brandy with water. It does not say how much water was allotted. Beef, mutton, pork or bacon are unmentioned. For each horse eighteen pounds of hay and one peck of oats with water At the end of the voyage all residue of provisions are to be strictly returned to store. If bread was not to be obtained, then one pound of biscuit per man was substituted. On shore at about the same time the dietary was somewhat different.

Every dragoon had 1½ lbs. avoirdupois 'of good and wholesome well baked wheaten bread.' This was very well if the dragoon got it—but did he? The bread he had to fetch from the bakehouse, and his allowance was 5½d. a day to pay for it. Bread must have been dear, as the quantity would be about three-quarters of an ordinary loaf in these days. He was allowed ½d. a day for fire, pickets, and other necessaries.

His horse had two trusses 'of very good old hay of the last year, weighing 56 lbs. each truss, and one truss of straw of 36 lbs., per week, to be fetched Monday and Thursday at a distance not exceeding two miles from the Camp at a rate of 6d. a man a night.' As the camp referred to was at Hounslow, this entry is of interest to the 7th Hussars. Later the scale was somewhat changed, and the paper from which it was derived is worth quoting.

For dragoons for seven days, two trusses of new hay of 64 lbs. a truss and one truss of 36 lbs. straw at 4d. a man a night. Also a stone of very good and sweet oats at 6d. a peck. These were to be fetched a distance also not exceeding two miles from the camp. Each tent was to be provided with two trusses of straw, new every ten days. Sea coal was to cost 7d. a bushel. Stakes or pickets were to be provided by Government. A brewer, one Captain John Grant, was under a bond to supply good sound wholesome beer to be sold at an agreed rate.

The sutlers were under bond to sell only this Government beer and to sell nothing but what is good. They had also to fill up the latrines when the camp was struck. Sutlers were to pay no licence fee, but their bond required a 12d stamp.

The Commissary-General is enjoined often to visit the sutlers in the field and see if all is good and safe which they sell.

Mention of the sanitary arrangements above leads naturally In the consideration of the hospital arrangements of that date. Of these the rules and regulations have come down to us in some detail; and a comparison with the existing hospitals is of interest. In each division of a military hospital—and these were founded in 1686—there was a matron and one chief nurse, who were in charge of the beds and furniture; the pay of the matron being £10 annually. During the summer only, a maidservant and cook were provided for sick soldiers at £3 each. Why in the summer only is not explained. Three nurses or tenders engaged by the matron had the care of the sick, washed for them, and watched them, at 5s. per week each.

A clerk was employed to keep the accounts. £10 for summer subsistence money for each soldier in hospital was paid to the matron, who was bound to feed the sick as the physician ordered, and also to find fire, soap, and candles. The Physician General or Chirurgeon (surgeon) General or Surgeon (if of a regiment) was to see each sick man daily and sign a report. Cases of sickness were to be promptly notified by the Commissary-General to the Paymaster-General in order that the subsistence money for the invalid should be duly paid. A printed certificate of discharge from hospital was ordered to be given to each sick man either by the captain of his troop or company or by the Provost Marshal. A sentinel was to be posted always to prevent the sick from rambling, or the healthful from coming to them without permission.

Until the year 1691 the supply of drugs for the army had been in the hands of the surgeons of the regiments, who purchased what they pleased, where they pleased, and probably were not too particular as to the quality of the stores they bought. In this year four very prominent physicians—Sir Thomas Millington, Dr. Hutton, Dr. Harald, and Dr. Lawrence—were called in to devise measures for supplying the army with good and pure drugs.

Now in England the manufacture and supply of drugs had since 1617 been vested by charter in the Society of Apothecaries of London. Of course this was a monopoly, but unlike most monopolies it worked well by preventing grocers and quacks from selling poisonous rubbish or useless confections. But between the physicians and the apothecaries there was a feud of long standing: it had already led to a Star Chamber cause. Hence, instead of going to the Society of Apothecaries as a society, to supply the army drug-chests, the physicians entered into negotiations with Jonathan Leigh, Colonel Robert Gower, Edwin Harle, and James Anderson, of London, Apothecaries. Gower had been a Court nominee for the post of assistant on the Court of the Society. Harle was the City sword-bearer, and in virtue of his office was elected to the Court, but declined to serve. Leigh or Lee was later the upper warden of the Society. Anderson cannot be traced as James, but if his name was William he was the gardener to the Society, and managed the Botanical Physic Garden; otherwise he was merely an apothecary and held no office. Clearly it was a breach of the charter of the Society for any four members to unite to sell to the Government to the detriment of the Society as a whole. But this they did. It was agreed that the Government should hand over three rooms in the Savoy to be used as an Elaboratory for the manufacture and store of drugs for the army. Two of these rooms were occupied by a man named Allen, a messenger, and the third was used as a prison for Irish prisoners. Chests of drugs, linen, vessels, and other things were to be supplied at rates agreed upon with the physicians, and these rates were £3 6s. 10d. each.

The apothecaries, of whom three out of the four had to proceed to Rotterdam, Brussels, Breda, Maestricht, and elsewhere, were to be provided with conduct by sea and horses on land, or in the case of horses an equivalent if desired of 7s. apiece *per diem*.

They are styled '3 able and careful apothecaries,' and are charged to deliver the chests of drugs to the surgeon of each regiment. And; it may be mentioned that the first known surgeon of Cunningham's Dragoons was one Peter Telfer (or Talifer), who by repute was singularly skilful in his profession.

The cost of drugs, &c, for the army in Flanders for the campaign of 1697 was only £3885; '95 pairs of chests of drugs and medicines and utensils and for so many battalions of foot and squadrons of horse at £23 a pair, £2185 0s. 0d.' The price of a chest of drugs had risen. But there are 'Magazines or stores of medicines and for the hospitals to be lodg'd in the towns in Flanders over and above what remains of the last year's stores—£1700 0s 0d.' The total is £3885.

To the original four physicians is given the task of examining into the purity of the drugs, and to them are added a Mr. Van Loon, a Mr. Rottermant, and others. The status of these gentlemen is unknown; but they were not on the Court of the Apothecaries' Society. In January 1702 another breach of the Apothecaries' Charter took place.

Sir Thomas Millington, apparently dissatisfied with the drugs supplied by Harle, Gower & Co., obtained from Queen Anne a letter directing the Bishopsgate Dispensary, a very early quasi-benevolent medical society, to provide medicines for the army and also for the fleet then going to the East Indies. The Society protested in vain. Later they obtained the contract for supplying the navy, and this they did for long years. But with, regard to supplying the army and the Army Medical Board when formed, no further negotiations went on till 1810. when the Board approached the Society.

Pages of letters and long lists of questions on one side and answers on the other passed between the Board and the Society. Could the Society supply drugs? Would they keep enough ready packed for home and foreign use? Would they establish depôts at Plymouth, Falmouth, Portsmouth, and elsewhere? Would they take back damaged drugs? The Society replied that they could supply the drugs and would keep sufficient stock, but could not establish depôts; and as for taking back damaged drugs, such would be of no use to them. The Government urged that other customers could have them. The Society suggested that if pure drugs of proper strength were needed, they were as equally needed for civilians as for soldiers. Then the Army Medical Board dropped these queries and put others. What profit, if any, did the Society expect to make? This was the outcome of a natural desire by the Society to know who was to pay them, how they were to be paid, and at what periods.

The Army Medical Board then suggested that as the Society already supplied the navy they might, if supplying both services, do it at a cheaper rate. To this no answer was returned. A short time elapsed. and then from the Army Medical Board came a long disquisition on the properties, strengths, and qualities of ipecacuanha, jalap, and Peruvian bark, which were apparently the three drugs principally used!

Their questions were answered, though a most unwarranted of adulteration of bark had been made by the Army Medical Board, and in a most offensive way.

With this the matter dropped, and it was not until ten years later that the secret of the failure of negotiations crept out. Mr. Garnier, the King's Apothecary, died in 1820, and it appears he had held, if not the actual contract all the time, at least such vested interests in the supply of drugs that he was master of the situation, and that situation was somewhat Gilbertian, for, being King's Apothecary, he was *ex officio* a member of the Court of Assistants of the Society of Apothecaries, and had been doubtless present at the sittings of the Court of Assistants, and quietly chuckling throughout the discussions raised there by the Army Medical Board's letters, suggestions, and questions.

But these early records of the Army Medical Service are not without, interest, being certainly, from one point of view, instructive, and, besides that, not entirely bereft of humour. It is not generally known that in the middle of the last century the detonating powder for the percussion caps of muskets, rifles, and pistols was made by the Society of Apothecaries for the Government at their laboratory. This laboratory still exists, and is still in use. It is attached to the rear wall of the Hall of the Society in Water Lane, Blackfriars. It will be admitted that this was a most extraordinary place to select for the manufacture of a most powerful and dangerous explosive. The manufacture there ceased after an explosion in which one of the operatives was blown to pieces. Luckily, only a very small amount exploded; as a matter of fact, the powder was a small experimental supply, which was being tested before a larger quantity was put in hand. Had there been even a few pounds only there at the time of the accident the havoc would have been tremendous. Things are, luckily, managed differently now.

CHAPTER 4

Highland Rebellion
1713-1716

The wholesale disbandment and reduction of regiments which immediately preceded the signing of the Treaty of Utrecht, 11 April 1713, and the similar proceeding widen, in the case of Ker's Dragoons and other regiments, followed it, had most dangerously weakened the strength of the British army.

The disbandment and reduction had been the outcome of three distinct political forces; one a professed desire for economy, the second the rooted objection which was popularly held to a standing army, and the third, and perhaps the most potent, the intrigues of the Jacobite party. All these factors were, of course, political.

Queen Anne died on 1 August 1714, and was apparently succeeded peacefully by George I. But this peaceful succession was more apparent than real, for throughout the Jacobites meditated a recourse to violence. Every possible engine of political intrigue that could be brought to bear to cause the disbandment of regiments of known Protestant, and therefore of Hanoverian sympathies, was exerted by the Jacobite party to the utmost.

Naturally Ker's Dragoons, being mainly Scots and the rest English, were Protestant. That some of their officers were or had been suspected Jacobite leanings is true, but it had been merely the malicious gossip of a court spy who, for a living, had to retail as much as he could to justify his miserable existence. Undoubtedly, on the accession of George I the security of the kingdom;—was in a somewhat parlous state. Of course, the hopes of the Jacobites naturally rested on a weak army; against a strong one they knew they could make no headway.

George I arrived in England from Hanover on 17 September 1714. Almost immediately it became apparent that if he was to retain his new kingdom a considerable increase to its military strength must be made, and measures to increase that military strength were speedily undertaken. Ker's Dragoons, which after a few months in Ireland as a dismounted regiment had been broken in the spring of 1714, were ordered to be re-established.

Now, while Ker's Dragoons had been disbanded, the Royal Dragoons and the Scots Greys had been augmented, and their increase of strength came from the disbanded men of Ker's Dragoons. Whether these men (five troops) were first discharged and then re-enlisted, or whether the five troops, two to the Royal Dragoons and three to the Scots Greys, were simply transferred to those regiments, does not appear. Neither do we know how the regiment had been employed while dismounted in Ireland, though probably police and preventive work was their portion;—it usually was the occupation of troops in Ireland at that time, varied only by occasional rebel hunts when there chanced to be rebels to harry. The officers were all placed on half-pay. George I determined early in 1715 to re-establish Ker's Dragoons, and the warrant so to do is here given.

George R.

Whereas we have thought fit that a regiment of dragoons be immediately formed to be under your command, to consist of one colonel, one lieut.-colonel, one major, one chaplain, one chirurgeon, and six troops, each consisting of one captain, one lieutenant, one cornet, one quarter-master, one serjeant, two corporals, one drummer, one hautboy, and thirty private dragoons, (including two for widows.) And whereas we have directed our right trusty and right well-beloved cousin Thomas, Earl of Strafford, to deliver over unto you the two youngest captains, two youngest lieutenants, two youngest cornets, and the two youngest quarter-masters, together with the non-commissioned officers and private men of the two youngest troops of our royal regimen of dragoons under his command, with the horses, arms, clothing and accoutrements; and also our right trusty and right well beloved cousin David, Earl of Portmore, to deliver unto you the three youngest captains, three youngest lieutenants, three youngest cornets, and three youngest quarter-masters, together with the non-commissioned officers and private men of the three youngest troops of our regiment of dragoons under his command with their horses, arms, clothes and accoutrements; our will and pleasure is, that you receive from the said Earl of Strafford, and the said Earl of Portmore the commissioned and non-commissioned officers and private men directed to be delivered over unto you as aforesaid, towards forming the said regiment of dragoons. And we do authorize you, by the beat of drum, or otherwise, to raise so many volunteers as shall be wanting to complete and fill up the regiment to six troops, each consisting of the numbers aforesaid. And all magistrates, justices constables, and other of our officers whom it may concern, are required to be assisting unto you, in providing quarters, impressing carriage and otherwise, as there shall be occasion.

Given at our court at St. James', this 3rd day of February, 1714-15, in the first year of our reign.

By his Majesty's Command,

William Pulteney

To our trusty and well-beloved Colonel William Kerr.

The term Widows Men requires a brief explanation. Widows Men were imaginary privates, usually two for each troop. They existed on paper only, and pay was drawn for them as if their existence was real. They were borne as phantom soldiers on the strength of the regiment. The sum of money which accrued in this manner was applied to form a fund for the purpose of supplying pensions for the widows of deceased officers. A warrant on this subject was issued in the reign of George I and bears date 26 April 1717

In those days the adjutant of a regiment held a position quite different from that held by an adjutant later. His duties were more nearly analogues to those of a regimental serjeant-major in these. His commission, however, was signed by the King, and therein he was styled Gentleman. A quartermaster was also styled Gentleman in his commission This commission was not, however, signed by the King, but by the colonel of the regiment. It very frequently happened that a quartermaster after a very short service received the commission of a cornet in the cavalry or an ensign in the infantry. Sometimes a man who had held a commission of superior rank

70

in a former regiment became a quartermaster in another, dropping his previous commission *pro tem.*

The list of officers of the re-established regiment is as follows :

Colonel Kerr's Regiment of Dragoons

William Kerr	Colonel	31 Jan. 1714/15
James, Lord Torphichen	Lieut.-Col.	31 Jan. 1714/15
Matthew Steuart	Major	
Lewis Dollon	Captain	
Peter Renouard	Captain	
William Crawford	Captain	
George Dunbar	Captain	
James Livingston	Captain	31 Jan. 1714/15
David Ogilvie	Capt.-Lieut.	
Alexander Auchenleck	Lieutenant	
William Delavale	Lieutenant	
James Ogilvie	Lieutenant	
Samuel Southouse	Lieutenant	
Bernard Lustau	Lieutenant	15 June 1715
William Richardson	Cornet	
George Lauther	Cornet	
George Knox	Cornet	
James Agnew	Cornet	
Henry Carlisle	Cornet	
John Keat	Cornet	
Rev. James Ramsay	Chaplain	
William Johnson	Adjutant	
Patrick Telfer	Surgeon	

Of these officers Captains Dollon and Renouard were transferred to Ker's Dragoons from the Royal Dragoons, and Captains Crawford, Dunbar, and Livingston from the Royal Scots Greys. The Rev. James Ramsey was chaplain of the Earl of Portmore's Royal Regimen of Dragoons 4 May 1714. Samuel Southouse was a lieutenant in Lord Cobham's Dragoon 6 November 1712. But the transfer of youngest captains, lieutenants, and cornets never appears to have been fully carried out in the manner ordered in the warrant quoted above.

Captains Dollon and Renouard with their troops joined from the Royal Dragoons, and Captains Crawford, Dunbar, and Livingston similarly joined from the Scots Greys. Lieutenant Southouse came from the Royals, but no cornets. Cornets Lauder, Knox, and Agnew came from the Scots Greys and no lieutenants. Ramsay the chaplain belonged to the Royals. There is no trace of the quartermasters.

The Regiment now consisted of five troops, to which a sixth was at once added, the men requisite being raised in or near London. On re-establishment the Regiment was permitted to retain its former rank and standing in the army. By 19 February the Regiment was equipped and five troops had received orders to march The three troops which were transferred from the Scots Greys were then in Scotland; the two transferred from the Royals were in England at Macclesfield and Pontefract; while the newly raised troop, which was under the command of the colonel, was in London.

The route was as follows: one troop from Dunse to Barnard Castle; two troops from Kelso to Richmond (Yorkshire); one troop from Macclesfield to Bolton; one troop to remain at Pontefract. Colonel Ker's troop to march from London to Skipton (Yorks.). On 15 March the troop from Bolton removed to Bradford and Otley.

The designation of the Regiment was now about to be changed, and never since 1 August 1715 has it been officially known by its colonel's name. Unofficially, and probably for convenience, for some little time the old designation Ker's Dragoons was used in letters and in print. The document authorising this change is here given:

Whitehall
1st August, 1715
Gentlemen, His Majesty having been pleased to declare the Regiment of Dragoons whereof the Honourable William Kerr is Colonel to be Her Royal Highness the Princess of Wales's own Royal Regiment of Dragoons, I am to desire you will acquaint the Lord Townshend therewith, that a Commission may be accordingly prepared, constituting the said William Kerr Esquire Colonel of the said Regiment.
I am, Gentlemen, your most humble Servant.
William Pulteney
Secretary at War
The Secretaries to the Lord Townshend &c, &c, &c.

This paper is extracted from the Notification Book, vol. 1317, fo. 100. The names of the Princess of Wales were Wilhelmina Carolina. It should be noted that at this period the Regiment was styled Royal. The next paper is dated from Whitehall, 2 August 1715, and is a letter to Colonel Ker in which he is ordered to hold his regiment of dragoons 'in readiness to take the field,' and to give immediate notice to the officers 'to provide forthwith everything necessary in order to it.'

On 15 August the Regiment shifted its quarters: two troops from Skipton, Bradford and Otley to Manchester and Sawforth (Salford?); one troop from Pontefract to the same places; two troops from Richmond (Yorks) to Wolverhampton and one troop from Barnard Castle to Leeds.

The Jacobite Rebellion, known familiarly as the 15, was now about to break out in Scotland.

Colonel Ker next receives a letter from Whitehall, dated 31 August 1715, in which the writer is 'commanded to signify' to him His Majesty's pleasure that 'you hold the Regiment of Dragoons under your command in a readiness to march at an hour's warning.' It is added that General Carpenter and Cobham had received similar orders. Some order which was now given is missing, but in the Regimental Record we read, under date 3 September, that 'Notwithstanding formal orders, the Regiment to march into North Britain,' viz. One Col. Ker's troop from Skipton, one Captain Dollon's from Pontefract, one Captain Renouard's from Bradford, one Captain Crawford's from Barnard Castle, one Captain Livingston's from Richmond, and one Captain Dunbar's from Bedal. All six troops were to rendezvous at Stirling Camp, N. B.

Early in September the rebellion broke out The Earl of Mar raised the Pretender's standard in the Highlands, and to that standard flocked the disaffected clansmen. Meanwhile Major-General Whetham was encamped in command of the royal force then assembling or assembled at Stirling. It would appear from the following paper that orders had been issued for all half-pay officers to immediately repair to Stirling camp, but that for some reason this order was cancelled.

Whitehall
28 Sept., 1715
My Lord, Enclosed I send your Lordship a copy of his Majesty's

Warrant to M. G. Whetham in which his Majesty is pleased to countermand the orders for the half-pay officers of your Lordship's Regiment of Dragoons to repair to North Britain, of which public notice will be given in the Gazette.
I am, my Lord, your Lordship's &c.
William Pulteney
E. of Hyndford.
The like to Col. Kerr.

In contradiction to this, in one of the printed accounts of the campaign it is distinctly stated that Major-General Whetham made satisfactory arrangements for the half-pay officers at Stirling and in several other places, where they were placed in charge of small garrisons and outposts; so that presumably they went there after all, and probably a fresh order to that effect was issued, a copy of which is not now forthcoming. At the camp at Stirling the Regiment remained for some time.

In September orders were sent to Major-General Whetham, the then commander-in chief in Scotland, to secure the important pass of Stirling Bridge over the Forth. To several officers was given the charge of watching the coast, so as to prevent the landing either of the Pretender, suspected persons, arms, or ammunition. Nevertheless, one small vessel arrived at Arbroath which carried some gentlemen on board from France, and also a supply of munitions of war. A few days later another ship got through, carrying passengers who at once proceeded to Mar's rebel camp at Perth. They also announced the speedy arrival of the Pretender.

Thereupon, Mar resolved to march for Edinburgh, and purposed to cross the Forth, if possible, about five or six miles above Stirling. October 18 was appointed as the day for starting; the plot or rather the intention of the rebels leaked out, and finding a detachment of the royalists waiting upon his march, Mar abandoned his project after proceeding some miles and returned to Perth.

Argyll now arrived and assumed the chief command in Scotland according to his commission. En route to Stirling he summoned Colonel Ker and Major-General Whetham to Edinburgh, where he left them in command of 100 dragoons and 150 foot, with some militia and volunteers, to protect the city and also to carry on the siege of Seaton House. No great garrison this, it must be admitted, but men were scarce. The Duke of Argyll left Edinburgh on 16 October and arrived at Stirling on the next day. He took with him 200 cavalry and 50 foot.

The strength of Ker's Dragoons on 29 October 1715 was '6 serjeants, 6 Drummers, 3 sick, 10 Entertained, 0 in prison, 153 Corporals & Centls Effective, Total of Corporals & Centls 166, 0 Dead, 0 Deserted.' This is extracted from State Papers, Scotland, Domestic, Bundle 9, No 92. The meaning of Entertained is probably recruits as yet unfit for the ranks. Total 153+3+10=166. Centls stands for Sentinels and has the same signification as private men, troopers, privates.

Mar now issued a proclamation which had for its intention the raising of money, for ready money was scarce in the rebel camp. Argyll, who was now commander-in-chief of the royal army at Stirling, upon 25 October issued a counter-proclamation. Two days later he issued another, this time for recruiting purposes. On 1 November Mar replied to Argyll's recruiting proclamation by a counter-proclamation.

On 20 October news reached Edinburgh that some 2300 men of the Western Highland clans under command of General Gordon had threatened Inverary. They approached within half a mile of the place, but finding the garrison ready to receive them, drew off. Here they received a reinforcement of 300 of the Karl of Breadalbane's men, and the rebels, now 2600 strong, marched off to join Mar at Perth. The preservation of Inverary was a very important matter. The commandant there was the Earl of Islay, brother to the Duke of Argyll.

On 23 October Argyll at Stirling received news that a party of rebels, 200 foot and 100 horse, were marching towards Dunfermline via Castle Campbell. A detachment of dragoons under Colonel Cathcart (Scots Greys) were sent after them. The royalists came up with the enemy on the 24th at 5a.m. They attacked them with vigour, killed and wounded several, and returned to camp that evening with seventeen prisoners, mostly gentlemen.

A few days later a detachment sent by the Earl of Hay surrounded and dispersed a body of 400 rebels (Breadalbane's men). But Mar's force all this time had been growing in numbers, and now amounted to over 8000 men. He therefore prepared to march from Perth to join General Gordon with the western clansmen at Auchterarder, in order to attempt the passage of the Forth. This was on 12 November. Argyll, however, heard of the movement, and in all haste ordered up a train of field artillery from Edinburgh. He could expect to receive no other reinforcements and determined to give battle before the rebels could reach the bank of the Forth. Accordingly he passed the river by Stirling Bridge and advanced towards Dunblane.

On Sunday 13 November was fought the battle known as Sheriffmuir or Dunblane. Of the engagement there are at least four printed contemporary accounts, besides many manuscript letters which contain either professed narrations of the events or references thereto.

Let us endeavour to disentangle the variations in these accounts, a matter of no little difficulty, since both sides claimed the victory.

Colonel Harrison, who was sent express to King George by the Duke of Argyll immediately after the battle, and who arrived at St. James's on Saturday 19 November, gives an excellent account of the disposition of the royal and rebel armies. Argyll, he states, chose Dunblane because it was much more advantageous for his horse than a position at the head of the river Forth. He had ascertained that the rebels designed to encamp at Dunblane that night. Hither then he marched on the evening of 12 November, and encamped with his left at Dunblane and his right towards Sheriffmuir. The enemy stopped short that night at a spot some two miles distant.

Next morning Argyll heard from his advance guard that the rebels were forming and rode to a rising ground to reconnoitre. Here he had a good view of the enemy and observed that his right flank was threatened. The moor on the right had been impassable on the previous night, but frost had hardened the surface and it was now possible for troops to cross it.

Argyll, therefore, extended his troops on the right in the following formation:

> Three squadrons of dragoons upon right and left in the front line, & six battalions of foot in the centre. The second Line was composed of two battalions in the centre, one squadron on the right, and another on their left, and one squadron of dragoons behind each wing of horse in the first line.

76

But it must be remembered that his whole force was barely 3000 men. When the right of the royal army came over against the left of the rebels, which rested on a morass, Argyll perceived the enemy was not quite formed and gave an order for an immediate charge. This was carried out, and the royalists 'charged both their horse & foot,' being received ' very briskly.'

After some resistance the royal troops were not to be denied; they broke through the rebel left and put them to flight. The retreating clansmen were pursued for some two miles by dragoons, infantry, and a squadron of volunteer horse.

The pursuit was continued as far as the river Allan, and the royalist troops deemed that the whole rebel army was discomfited and in retreat. It is on record that the enemy several times attempted to rally and to make a stand, but each time they were attacked and broken.

Having arrived at the river Allan, Major-General Whetham, who was in command of the infantry, sent to the Duke of Argyll to inquire what had become of the troops forming the royalist left, as he could see no signs of them, and also that a considerable force of rebel horse and foot was formed in his rear.

Argyll halted, formed his troops in order, and marched towards a hill upon which the rebel horse and foot had by this time formed Argyll then extended his right towards Dunblane to give his left an opportunity of joining them. Awaiting the left his men remained then until it was quite dark, upon which they slowly withdrew to the ground on which they had formed in the morning,

The rebels on the top of the hill, still un-dispersed, thereupon moved off to Ardoch. About half an hour later 'our troops which had been separated from the Duke of Argyll joined his Grace.' This was, of course, Whetham's force. Meanwhile, what had befallen the royalist left? The story appears be this.

In the beginning of the action the dragoons on the left charged some of the rebel horse on the rebel right, and captured a standard. The rebel foot, however, attached the royalist foot with such vigour that the latter were thrown into complete disorder and were obliged to take refuge among the dragoons, the whole being mixed up in inextricable confusion. The effect of the rebel attack was to cut the royalist left completely off from the remainder of the royal army. A message or news of some kind at this period reached them to the effect that a body of the enemy was endeavouring to get to Stirling, and by great exertions their officers managed to rally and bring off their men so as to seize and occupy the passes leading there, thus, in a way, saving the situation.

But a hideous slaughter took place during the time that the royalist left was in confusion. In a certain sense, therefore, both sides could claim the victory. To the Duke of Argyll's army the spoils of war were 'fourteen colours and standards, four pieces of cannon, tumbrels with ammunition, and all their bread-wagons.'

Viscount Strathallan, two colonels, two lieutenant-colonels, one major, nine captains, besides subalterns, were taken prisoners and lodged in Stirling. Lord Forfar, of whom we shall write later, was wounded, mortally, as it proved. The Earl of Islay, who had only arrived two hours before the battle, received two bullets, one in the arm and one in the side; General Evans was cut over the head; Colonel Hawley shot through the body; Colonel Lawrence taken prisoner; Major Hames and Captain Armstrong, the latter aide-de-camp to the Duke of Argyll, were both killed.

The reference to the wounded Lord Forfar are many, and are to be found mainly in the correspondence of the Duke of Argyll. These letters are preserved the Scottish State Papers, Domestic They form very interesting reading but cannot, of course, be quoted at length. A few extracts may however be given.

> The Lord Forfar was shot in the knee, & cut in the head, and receiving ten or twelve strokes after he had got his quarters (obtained quarter) from the rebels: He is in very great danger of his life. Of his Regiment Ensign Branch and eight private men killed.
>
> My Lord Forfar, who acted as Brigadier and charg'd at the head of Morrison's Regiment, was barbarously butchered.
>
> Of our side the Earl of Forfar is missing, and we're afraid is killed.
>
> The Lord Forfar lies dangerously ill of his wounds at Stirling. After he was shot threw the knee, he received sixteen wounds from their broad swords.

Dated from Stirling 26 November 1715. From the Duke of Argyll to Lord Townshend:

I was this night sent for by poor Lord Forfar who is a dying, who made me promise to pray the favour of Your Lordship to represent to his Majesty that the money with which he purchas'd his employments was lent him by his Aunt Mrs Lockhart, and that having behaved himself to the approbation of everybody, and dying in his Majesties Service, possessed of too small a fortune to repay her, He humbly hopes from His Majesty's great goodness that he will be pleased to permit the Regiment to be dispos'd of for four thousand pound to repay her.

The letter concludes with a recommendation of Lieutenant-Colonel Cathcart of Lord Portmore's Dragoons for gallantry at the battle, and suggests that that officer should be promoted Brigadier.

The expression 'permit the Regiment to be dispos'd of for four thousand pounds' needs brief explanation. It was not the regiment that was to be sold, but the colonel's commission to command it. This could be sold during the lifetime of the holder, but its value would be lost to his heirs if he died before the sale were completed. Lord Forfar was hourly expecting death, and it would seem had begged for a special act of grace from the King in reward for his services.

Stirling, 27 November 1715:

Poor Lord Forfar is still alive, but then is little or no hopes of his recovery.'

Two days later comes more favourable news in an Edinburgh letter from Andrew Hume to Pringle the Secretary:

They write from Stirling that the Doctors have good hopes of Forfar.

The hopes were illusory, for we next hear of his death. In a postscript to a letter from Argyll to Townshend, dated 10 December 1715, we read:

P. S. My Lord, Poor Lord Forfar died last Wednesday.

The amount of sympathy felt for the deceased soldier is very marked, and we can but surmise that he was a most popular as well as a most brave and gallant officer. Lieutenant-Colonel Cathcart was promoted Colonel 15 February 1717; Brigadier-General 27

November 1735. Apparently the recommendation of the Duke of Argyll did not carry sufficient weight.

The rebel force amounted to 9100 men, according to General Whetham's dispatch. The royal army 'did not consist of above a thousand dragoons and two thousand five hundred foot,' and but little more than half of were engaged. General Whetham writes of the enemy thus:

> I must do the enemy the justice, to say, I never saw Regular troops more exactly drawn up in line of battle, and that in a moment; and their officers behaved with all the gallantry imaginable.'

The Earl of Mar says:

> We discern'd a body of the enemy on the north of us, consisting mostly of the Grey Dragoons (Scots Greys), and some of the Black. . . . Our baggage and train horses had all run away in the beginning of the action. . . . Had our left and second line behaved as our right, and the rest of the first line did, our victory had been compleat: But another day is coming for that, and I hope e'er long too.

The day did not, however, come.

On 13 November General Wills and General Carpenter forced a complete surrender of the English Jacobites at Preston.

Writing on 16 November the Governor of Burnt Island, a Jacobite, writes on the authority of the Governor of Perth that '800 royalists fell and only sixty rebels,' and that '1500 stand of arms were captured from Argyll's men. A Jacobite account of the battle describes the failure of the rebel left to repulse the royalist right thus, and gives the following information:

> An incident happened, which contributed considerably to the Duke of Argyll's success:—one Drummond, an officer in Argyll's army, went to Perth, as a deserter, and communicated information to Lord Drummond, who made him his aide-de-camp. During the action of the 13th, the Earl of Mar, perceiving his right wing successful in repulsing the Duke of Argyll's left, despatched Drummond to General Hamilton, who commanded Mar's left, with orders to attack the enemy resolutely,

Instead of communicating these orders, he informed the General that the Earl of Mar, being defeated on the right wished him to fall back immediately, with as much order as circumstances would permit. General Hamilton, agreeable to these orders, gave way without firing a gun; when many of his men were killed, wounded and taken prisoners. Another incident aided the Duke of Argyll:—Rob Roy M'Gregor, noted few with resolution and courage, being within a little distance of the Earl of Mar's army with his men, when desired by one of his friends to go and assist, answered, "If they cannot do it without me, they shall not do it with me."

At the battle the Princess of Wales's Own Royal Dragoons were on the left of the royal line. When the rebel attack upon the royalist infantry threw the latter into confusion, Colonel Ker led his regiment to the charge in a most gallant manner. In this charge General Carpenter's squadrons and some gentlemen volunteers were associated. The rebel horse were completely routed and lost a standard. Colonel Ker had two horses killed under him and lost a third. A pistol bullet struck him in the breast, but, luckily, did not injure him, merely tearing his coat. The Regiment lost two troop horses killed, and one man and four horses wounded.

Then followed the discomfiture of the royalist foot, who were driven back and became mixed up with the dragoons as has been stated. Lord Torphichen, the lieutenant-colonel, distinguished himself greatly in the charge. On the morrow a body of dragoons went out to the battle-field and brought in the wounded to Stirling. The rebels accused the royalists of stripping their prisoners and of wounding them after quarter had been asked for and granted. An instance they gave was the case of Lord Panmure. The royalists made similar accusations, and cite Lord Forfar as an example.

Thirty years later the treatment of the wounded and prisoners by the Scots after Prestonpans was most humane.

Of the Duke of Argyll it is written that:

. . . . he behaved with great humanity and himself parried three strokes aimed by a dragoon at a rebel gentleman who was wounded and begged quarter.

Elsewhere we read, 'For a time the Dragoons gave no quarter.'

This was when they cut up the rebel cavalry on the right. Some of the comments of the Duke of Argyll are worth quoting. In a letter to Lord Townshend, dated from Stirling 21 November, he states: He is glad the re-enforcements of Dutch Troops are going to be sent.' He had been clamouring for them for sometime, knowing full well that the strength at his disposal was not sufficient to safe-guard the the kingdom. What would not have been his plight if, in lieu of going to Preston, the English Jacobites had turned north and hemmed him in between that army and Mar's?

Argyll adds to Lord Townshend 'that he has something to say in support of the need for them' (the Dutch). but he hopes 'your Lordship will think it very necessary to keep it secret, it is that some of our troops behaved as ill as ever any did in this world, which makes it the more wonderful that the day should have ended as it did.' This letter of Argyll's a good one.

Among other difficulties his troops needed subsistence, that is say, pay. It was due on 24 November, and he writes: 'I have not as yet heard of the Paymaster who was to come down.' He beseeches the Government not to be over-confident from the fact that his small army got rather the best of the battle: he says that s success of 3000 against 9000 was not likely to be repeated,' it having been due to luck, or at any rate to accident.

David Dalrymple, writing to the Lord Advocate, states that he encloses an 'extract of a letter from Lord Torphichen, Lieut.-Col. of Ker's Dragoons, which gives the plainest view of the action on our left that I have seen.' Unfortunately the extract is not contained in the letter. Of Lord Forfar and others Argyll writes to Lord Town-shend on 6 December:

> Poor Lord Forfar was at the head of Col. Morrison's who behav'd himself as well as a man could do, Major Hammers who commanded the Regt. was killed at the first fire. Lord Forfar soon fell and the Regt. Was as so overpowered that I wonder how Capt. Lloyd was able in those circumstances to bring off one of his colours and the remainder of the Regi-ment for which I think well deserves the major-ship.

This was Captain Leonard Lloyd, and he was promoted major in 1716, Argyll goes on to say that

Col. Kerr's two Squadrons & Lord Stairs Second Squadron, tho' they were carried off with the rest on that occasion are troops that may be depended upon, the Colonel and the two other Commanding Officers I am persuaded did their best and will do extremely well upon .ill occasions, and I do most sincerely think that what happened was Mr. Whetham's misfortune and not his fault.

The long looked for Dutch troops to the number of 1200 arrived on 15 December. Dated 24 December from Whitehall we have the following answer to a letter from Colonel Ker to the Duke of Marlborough, in which he appears to have asked for a commission for his adjutant. Reference is also made to Lord Torphichen and the Major of the Regiment.

Whitehall,
24 Dec. 1715
Sir, I am to acknowledge the favour of yours of ye 10th Inst., which I do with pleasure. I must acquaint, you that my Lord Marlborough is at present out of town, and that I shall not fail at his return speaking to him about ye commission for Adjutant Johnston as you desire, and then you shall hear further from me.
Be pleased to assure Lord Torphichen and your Major no opportunity that offers of doing them service shall be lost by
Sir, your most humble servant
William Pulteney
Coll. Kerr

The major here mentioned was James Nasmyth.

On 29 December the ship with artillery, which should have reached Stirling months before, was still in the Thames. However, the arrival of the 1200 Dutch troops eased the mind of Argyll.

On the last day of the year we read in a letter addressed from Stirling to Pringle:

Yesterday his Grace sent a detachment of 100 dragoons & 100 foot of the troops here to Dunfermline to remain there, and they re to be re-enforced by a detachment from some of the Dutch Regiments which quarter near the place, that is to say Queensferry & Borrowsbounes (sic).

83

From the Manuscript Regimental Record the following account of the battle is extracted:

> The Regiment was present at the Battle of Sheriffmuir. The order of battle for the cavalry was three squadrons on the right of the first line of infantry, namely Evans's, Scotch Greys, and Earl of Stairs, to the left these three other squadrons of dragoons, Carpenter's, Kerr's, and a squadron of Stairs under the command of General Whiteham (Whetham). In the second line of infantry a squadron of dragoons supported each extremity.
>
> The battle having terminated the trophies of victory remained with the Royalist General He had taken three standards including the Royalist [1] one called the Restoration. 13 pair of colours, 4 pieces of cannon, 7 wagons, and one silver trumpet. Nearly the whole wreck of the battle fell into his hands, including a great quantity of muskets, plaids, and broadswords of which many had silver cases for the hands. He had gained every advantage but that of being able to follow up the partial victory of the preceding day with a second attack, or with a pursuit, the numbers and condition of his men were quite inadequate or any such movement and while Mar fell back on Perth, he judged it expedient to return to Stirling.

An extract from The Flying Post, No. 3735, testifies to the gallantry of the Hon. Colonel Ker, but we need not quote it as its substance has already been given.

On 18 January 1716 additional men and horses which were being raised for the Regiment in England were ordered to be quartered at Dunstable and Newport Pagnall.

Meanwhile, the various troops of the Regiment had been scattered about in quarters in Scotland. The colonel's was at Leslie, Captain Dollon's at Dunfermline, Captain Renouard's at Dunfermline, Captain Crawford's at Kinross, Captain Livingston's at Cupar, and Captain Dunbar's at Anstruther. Here they remained until 27 March, when orders to march into England were received.

---

1. Of course the Royalist Standard mentioned was Rebel, and it is merely a mistake in the Regimental Manuscript.

Accordingly the various troops in the order named above marched to Ripon, Rotherham, Otley and Bradford, Selby and Cawood, Ripon, and Knaresborough.

Whether the Regiment accompanied the royal troops when an advance was made against the rebels in January 1716 does not appear, but considering the fact that even when reinforced the royalist army was not excessively strong, it is reasonable to suppose that the Regiment took part and share in the proceedings. The rebels fled. The rebellion as far as armed resistance being now at an end, the Regiment in consequence left Scotland for England, as has been already stated.

It is curious to note that in all the references to the Regiment, whether in manuscript or in print, the old style of 'Ker's Dragoons' is maintained. Even in the Manuscript Regimental Record it is so written. Can it be that the length of the designation 'Her Royal Highness the Princess of Wales's Own Royal Regiment of Dragoons' was found in practice to be somewhat cumbersome, or was it the force of habit?

CHAPTER 5

Home Service
1716-1742

We have already traced the history of the Regiment for the first twenty-six years of its existence as an integral portion of the British army. During that period it has twice been engaged in campaigns in Flanders and once in combating the Jacobite rebels at home. The period to which attention must now be given is one in which the Regiment was employed solely on home service. That the story of these next twenty-six years will prove interesting reading cannot for a moment be even suggested. Still, it is fitting that their record should be told with as much detail as possible, seeing that some fifteen pages of none too closely written folio sheets in the Manuscript Regimental Record contain the only detailed account of the period in question now existing.

Should this record, uninteresting as it may seem, be lost or destroyed, it could never be replaced. Hence for mere considerations of safety it is here set down and as far as possible amplified. Roughly speaking, the only events to be chronicled are changes of quarters in England, marches into Scotland, changes of quarters in Scotland, and an alteration in the style of the Regiment, which in 1727 was renamed the Queen's Own, a title it has retained until the present day. During this period too the death of the veteran Colonel, the Hon. William Ker, took place after a tenure of command which existed for nearly thirty-two years. Beyond these facts and mention here and there of reviews, royal escort duties, and a few cases in which detachments of the Regiment were called upon to assist the civil authorities in maintaining order there is little to tell.

On 14 May 1716 orders were received for Ker's Dragoons—the old title, be it observed, being still used—to change quarters. In consequence the troops of Colonel Ker and Captain Levingston left Ripon for Bedford and Newport Pagnall. Captain Crawford's, which had concentrated at Selby, proceeded to Ampthill. Captain Renouard's, which had concentrated at Bradford, and Captain Dollen's, then at Rotherham, marched to Bedford and Stony Stratford respectively; while the troop of Captain Dunbar left Knaresborough for Woburn.

On 4 June the two troops at Ampthill changed quarters for a few weeks and proceeded to Layton, returning prior to August. This Layton is probably either Leighton-Buzzard or Leighton in Huntingdonshire. During August the troops from Ampthill and Woburn proceeded to Ipswich; those from Newport Pagnall and Bedford to the same town, and those from Stony Stratford and Bedford to Bury St. Edmunds. On 25 November the troops at Bury St. Edmunds were moved, one to Sudbury and the other to Hadley and Stratford. This last move was made in consequence of an order from the Prince of Wales, who was then acting as regent during the absence of the King in Hanover. In the document the spelling of the colonel's name is correct, being 'Ker.'

Four days later the Ipswich troops proceeded to Woodbridge and Stowmarket. In another order referring to the same duty the Regiment is styled 'Her Royal Highness the Princess of Wales's Own Royal Dragoons.' It is sent by direction of the Prince of Wales and bears date 1716. By it one troop is ordered to be at Orford to attend the King as far as Woodbridge and one at Woodbridge for similar duty as far as Ipswich, and a third at Ipswich to proceed as escort either to Manningtree or Colchester.

On 17 December the Prince of Wales issued another order in which the same style is used, by which the quarters of the Stowmarket troop were enlarged to Needham. A paper in the Record Office is here worth quoting as far as it concern the Regiment. It would appear that there were certain arrears of forage money due to the Regiment—and also to other dragoon regiments during their stay in Scotland. This was an allowance of forage money over and above the ordinary pay which had been granted by Queen Anne. Colonel Ker and the colonels of the other regiments concerned petitioned the right Hon. George Ireby, Secretary at War, to 'move the King to grant it.'

Colonel Ker's account we here give. It is to be found in the Record Office and bears date 24 December 1716

State of the additional forage money due to Her Royal Highness the Princess of Wales's Own Royal Regiment of Dragoons commanded by a Honourable Col. William Ker.

One troop	Horses	
28 private men ye 2 widows' men exact	28	
1 serjeant	1	
2 corporals	2	
1 drummer	1	
1 hautboy	1	
3 captain's servants	3	
2 lieutenant's servants	2	
2 cornets servants	2	
1 quartermaster servant	1	
One troop	41	
	x6	
Six troops	246	
One to the adjutant, one to the surgeon	2	
	248	
248 horses at 1d. each *per diem* from 1 to 24 Oct. 1715 both included, being 24 days at £1.0.8 *per diem* is for ye whole		£24 16 0
Horses as above	248	
Additional	60	
308 horses at 1d. each *per diem* from 25 Oct. 1715 to 24 Dec. following inclusive being 61 days at £1 5 0 *per diem* for the whole		£78 5 8
6 horses more added, being 1 of the widows' men taken off the establishment for widows & added to the troop	6	
314 horses at 1d. each per diem from 25 Dec. 1715 to 30th April following inclusive, being 128 days at £1. 6. 2 per day is for the whole		£167 9 4
		£270 11 0

Will Ker

The arrears were eventually paid. Of course the paper in the Record Office is a copy, but it is interesting to see that whereas the colonel name is in most documents written 'Kerr' he himself signs 'Ker.'

On Christmas day the Stowmarket troop was ordered to march to Beccles to attend His Majesty 'in case he should land there,' and attend him to Saxmundham and afterwards to return to their quarter Apparently therefore, the King did not land at Orford on the day originally fixed. We can imagine that the men in general hardly welcomed this particular duty at the so-called festive season. On the same day the Orford troop was sent to Saxmundham. Beccles, by the way, is at least ten miles inland, up the river Waveney, at the mouth which Lowestoft now stands and did then. On 4 January 1717 half the troop at Saxmundham was sent to Kelswell (probably Kelsale near Saxmundham) and to Yoxford, there to remain until further orders.

February 6 a troop was ordered to march from 'Stowmarket and Needham' to Beccles. How half or a portion of this troop got to Needham (Needham Market) is not stated. March 18. The Woodbridge troop proceeded to Halesworth. On 4 April all the troops were shifted, two to Colchester, one each to Ipswich, Braintree and Sudbury, while one was divided between Hadley and Stratford,

Here the Regiment remained until 16 May, when the troops were moved to the following places: Sudbury to Peterborough, Hadley and Stratford to Peterborough, Colchester to Wisbech, Braintree to Wisbech, Ipswich to the Three Deepings (Market Deeping near Peterborough, Deeping St. James near the same city, and Deeping St. Nicholas near Spalding) The sixth troop went from Colchester to Spalding.

On 1 June the troops at Spalding and Wisbech were sent to Grantham and Dunnington respectively. Dunnington is evidently intended for either Donnington near Spalding or Donington-on-Bain, Lincoln There are two Donningtons, one near Alcester in Warwickshire: and another in Yorkshire: obviously it was neither of these. During this month a detachment of each of the troops from Peterborough, Wisbech. the Three Deepings, and Spalding were dispatched to Lydd in Kent, but for what purpose the Record does not state. Probably, however, it was for revenue duty. At Lydd we read that they relieved a detachment of 'Brig, Gore's Dragoons.'

On 27 June, as the Annual Fair and Horse Races were being held at Peterborough, the two troops at that place were ordered to march to Stamford until the festival was over, when they were to return.

From the entry under date 'July 9' we learn that the troop at Grantham and Spalding was commanded by Captain Dunbar, for it was it that day ordered to march to Holbeach and Fleet, the two troops at Stamford being ordered to remain where they were then quartered.

The Regiment was in October moved into Yorkshire. On 19 October the troops at Wisbech, Holbeach, Donnington, the Deepings, and Stamford were sent to Halifax, Bradford, Sheffield, Chesterfield and Doncaster, the detachment at Lydd being ordered to march thence on 25 October and to join their respective troops at the above-named quarters.

November 5. 1717. The Regiment was now very much weakened in numbers, as by an order dated from Hampton Court it was to be reduced to '25 private men per troop and no more' (Record Office).

The last move that year took place on 21 December, when the Sheffield troop was sent to Barnsley. For the year 1718 the events are follows: on 25 February one Doncaster troop marched to Rotherham. Two days later the Halifax troop marched to Skipton. On 3 March the other Doncaster troop and the Skipton troop moved to Leicester; lastly, on 5 April, the Barnsley, Rotherham, Chesterfield and Bradford troops proceeded to Leicester, Lutterworth (2) and Loughborough respectively. Later in the month, that is on 28 April, one of the Lutterworth troops was transferred to quarters at Ashby-de-la-Zouch.

Dated Whitehall, 31 May 1718, is an order directing one serjeant and six private men from the troop at Leicester to march thence to Prasby and to remain there as a guard upon the horses at grass. This information is derived from a paper in the Record Office and is the earliest mention of the horses of the Regiment being at grass quarters, the Regimental Record does not allude to grass quarters until later. Reference to the P.O. Directory fails to identify Prasby.

In its present situation the Regiment was permitted to remain in peace until 10 July, when orders were received to concentrate at Leicester, there to remain until reviewed by Major-General Macartney. After the review all troops returned to their quarters.

Orders to hold themselves in readiness for review were on the same date issued to Honywood's, Molesworth's, Gore's, Stanhope's and Cobham's regiments of dragoons.

On 17 July 1718 an order was received for the troops at Leicester to leave that town three clear days before the assizes were held.November 15 1718. It appears that there had been a mutiny in Molesworth's Dragoons, and that six mutineers were detained at Nottingham. Here they were delivered over to a detachment of Wade's Horse who escorted them to Leicester. At Leicester Ker's Dragoons had orders to take them over and escort them safely to Northampton, where they were to be delivered over to the commanding officer of Lord Irvin's regiment of horse. He had to transmit them to St. Albans, where they were finally handed over to a detachment of the foot guards, who took the unfortunate men to London.

On 6 January 1718-19 the detachment of 'Ker's Dragoons at Lydd' was ordered to march to Leicester, Loughborough, Lutterworth and Ashby-de-la-Zouch, to rejoin its various troops. There is not, however, any record of this detachment having been sent thither. It may, however, have relieved the one already withdrawn 25 October 1716.

On 5 March the troops at Loughborough, Ashby-de-la-Zouch, and Lutterworth proceeded to Wickware (Wickwar, Glouc.), Chipping Sodbury and Marshfield (near Chippenham) respectively, while the three Leicester troops moved to Malmesbury and Wooten underedge (Wootton Bassett? near Swindon). Two days later the route of the four troops who were ordered to Wooten underedge, Malmesbury and Marshfield was suddenly changed, and they were directed to proceed to Cirencester. There they were to halt till further orders. The route of the other two troops was also changed, and they were sent to Tetbury and Malmesbury, in which places they too were to halt. Here these two troops remained until 25 March, when they were directed to join the rest of the Regiment at Cirencester.

On 4 April two troops were sent to Burford (Oxon), and on the same date another troop was dispatched to Fairford (Glouc.).

This troop, from a later entry, appears to have been Colonel Ker's own troop, as it is so designated in an order dated April 17, when it was directed to march from Fairford to Lechlade (Glouc.) and there to await further orders.

On 21 April the troops at Cirencester, Lechlade and Burford

were sent to 'Hexham and places adjacent.' Hexham is a very ancient town in Northumberland. Here they remained until 20 June, when the whole Regiment was dispatched forthwith to Darlington (Durham), Yarum (Yarm?, Yorks, 4 miles south of Stockton), Northallerton (Yorks), Barnard Castle (Durham), Stockton (Stockton-on-Tees, Durham) and Bishop's Aukland (Bishop Auckland, Durham). On 4 July the two troops at Yarm and Northallerton were dispatched haste to Halifax to assist: the civil power in suppressing the riots in at neighbourhood. This was the first time in which any portion of the Regiment had been employed upon this duty—a duty, be it observed, which has never been popular in the service; still, being a duty, it has to be done, and has therefore, been performed.

For the civil authorities to call in the assistance of the military forces of the country to keep order when trouble was expected, or to put an end to disturbances when disorder had broken out, was in those days of far more frequent occurrence than happily it is in these. It may be asked why this should have been. The answer is simple. The civil authorities had no civil power at their backs to preserve the peace when broken, or to prevent a breach of the peace when it was either threatened or had actually taken place, or was believed to be about to take place. Those were the days when there was no police force in existence. Poverty and distress, owing mainly to the wars but lately concluded, were extreme, and this was one cause for riot. Private enmities against individuals or certain classes of individuals furnished other causes, the price of provisions another. Industrial disputes then, as now, however, were the most potent in provoking the more reckless to take forcible measures to redress what they rightly or wrongly considered their grievances. Lastly, there was then, as now, a proportion of the population ever ready to enter on a campaign of destruction, for therein lay opportunities for not only paying off old scores where personal animosity was a factor, but—and this was the main inducement—there was loot. The order dispatching the two troops to Halifax is in the Record Office.

Whitehall
4th July 1719
It is the Lords Justices' directions that you cause the two troops from Yarum and Northallerton to go to Halifax immediately and remain until further orders, and be assisting to

the Civil Magistrates to suppress the rioters in and about that place and to repel force with force in case it shall be found necessary to preserve the Public Peace.

Signed

George Ireby

Search in old newspapers has not furnished any details of these riots. County Histories are also silent on the subject. By 28 July, order having been restored at Halifax, the two troops had returned to their quarters.

On 19 August the Northallerton troops marched to Bedale (Yorks). On 9 October the three troops at Stockton, Barnard Castle and Darlington marched to Preston, while the troop at Bishop Auckland and that at Bedale proceeded to Ormskirk (Lancs.), and the Yarm troop went to Earstang. Earstang evidently should be Garstang (Lancs.). This troop on 20 December was removed to Bolton. On 5 April 1720 the three troops from Ormskirk and Bolton proceeded to Lancaster and Kendal and three from Preston to the same places. Four troops were quartered at Lancaster and the remaining two at Kendal. On the same date from a paper in the Record Office we find that the horses were at grass at Hornby (Lancs.), as an order was issued for thirty men with a due proportion of officers to proceed thither as a guard.

In these places the Regiment remained until Nov. 1, when Ker' Dragoons—the old title is even yet used in the Manuscript Regimental Record—proceeded to Preston (3), Lancaster (2)—these last presumably the Kendal troops—and Ormskirk (1).

During its occupancy of these quarters on 3 December the two troops of the Regiment at Lancaster were suddenly dispatched on a curious duty—this was no less than to patrol the Lancashire coast in order to prevent persons from landing there from the Isle of Man. The Record thus gives the information: 'The two troops of Kerr's Dragoons at Lancaster to be aiding and assisting to prevent boats with persons coming over from the Isle of Man, to land upon any part of the coast of Lancashire in regard that there is advice that the plague is in that island, and to send notice to the 3 troops at Preston to give the like assistance and to repel force with force if necessary.' Evidently the authorities well very wisely determined to take no risks. The grounds for their apprehension of serious danger were ample, seeing that during that year (1720) no less than 60,000 persons had perished of the plague in Marseilles alone, the scourge was believed to have been brought to that city in a ship from the Levant. The scare, if scare it was, did not last long, for on 20 December one of the Lancaster troops was sent to Wigan.

For a year now the Regiment was ordered to Scotland. On 5 April 1721 three troops from Preston and the other three who were then at Ormskirk, Wigan, and Lancaster were sent to Berwick, 'where they will receive orders from the Commander-in-Chief in North Britain, for their further proceeding.'

We hear no more of the Regiment until 14 April 1722, when two troops were at each of the following places: Jedburgh, Dunse and Kelso. They had occupied the same quarters years previously when under the command of Lord Jedburgh. On that date they were all ordered to proceed forthwith to Wakefield, Pontefract and Doncaster.

On 8 May the Regiment was directed to concentrate at Manchester and encamp in the neighbourhood. In this camp they remained until 25 September, when they went into quarters in the town. On 13 October the Regiment marched from Manchester, three troops proceeding to Derby and three to Nottingham. Here the Regiment remained until 13 April 1723, when the Derby troops proceeded to York and the Nottingham troops to Derby.

On 11 June the whole Regiment was concentrated at York, and on that date its strength consisted of 333 men. As in the preceding year, the Regiment was sent into camp—this time near York. Here they remained from 20 June to 21 September, when they returned to the city. On 30 September the Regiment received orders to march, half to Reading and half to Newbury and Speenhamland on the outskirts of that place. The last entry for this year is dated November 30, and informs us that on that day:

Twenty men of Kerr's Dragoons to march by turns from Newbury to Reading for the convenience of the riding house built there.

This is quite an early mention of a cavalry riding school, at any rate of a covered one.

Early in 1724 a detachment of the Regiment was ordered on preventive duty. The entry referring to this is as follows:

6 Jan. Three men from each of the 6 troops of Kerr's Dragoons to march to Gosport and afterwards to be disposed of as follows:

4 men Parish of Kingston   4 men Parish of Fareham
4 men Parish of Tichfield   4 men Parish of Gosport

To be aiding and assisting upon the coasts of Hampshire and Dorsetshire when required by the officers of the Customs against Owlers and Smugglers.

An owler was one who was guilty of the offence of carrying either wool or sheep out of the country. It was punishable by fine or banishment. The wool trade was the staple trade of England in old times, but the only symbolical relic of it is the woolsack in these days. Tom Brown, who only died in 1707, has left us the following on this subject:

To gibbets and gallows your owlers advance,
That, that's the sure way to mortify France,
For Monsieur our nature will always be gulling
While you take such care to supply him with woollen.

Brown libelled the French King after the Peace of Ryswick and was put in gaol till by a whimsical rhyming petition he obtained his liberty. A quatrain often quoted is his—'I do not like thee, Doctor Fell,' etc.. The smuggler was one who brought or endeavoured to bring foreign goods into the country without paying the customs duty thereon.

The preventive detachment was not relieved until 14 April 1725, when a similar detachment was sent to take up the duty, and had to march all the way from Yorkshire and Durham for the purpose.

The new year found the Regiment below strength, and an order was issued to Colonel Ker on 20 January 1724-25 as:

Colonel of our most dear daughter Wilhelmina Carolina Princess of Wales' Royal Regiment of Dragoons' by beat of drum or otherwise to raise volunteers in any 'County or part of our Kingdom of Great Britain.

It will be remembered that recruiting in Ireland was absolutely forbidden at that time.

April 3, 1724. The Regiment was ordered to march from Berkshire Yorkshire and Durham; its appointed quarters being Ripon (two troops) and Richmond, Darlington, Knaresborough and Bishop Auckland (one each). On 16 May the Bishop Auckland troop was moved to Yarm and Stocksley (Stokesley, Yorks., twelve miles south of Middlesbrough).

One of the troops was upon 17 September sent to join the two already at Ripon and the other three troops were concentrated at Leeds, where they were ordered to remain until reviewed, after which they were to return to their former quarters. A month later the Yarm troop was sent to Tadcaster. There is apparently a gap in the records here, as the Tadcaster troop would seem to have left that place prior to 1 May 1725. On that date two troops of the Regiment were ordered to march 1 troop from Richmond to Tadcaster, 1 troop from Darlington to Selby (Yorks.) and Cawood (Yorks.), while one was to remain at Knaresborough and three at York.

On 2 September the whole Regiment concentrated again at York, where they were reviewed by Lieutenant-General Carpenter, after which they returned to their various quarters.

The last entry for 1725 tells us that the three troops at York were sent two to Wakefield and one to Pontefract. It is dated 7 Oct.

The events of the year 1726 are as follows. The first, dated 19 February, is somewhat obscure. 'The three troops at York to remain during the assizes and afterwards return.' The second, dated 5 April, is equally so. 'The 3 troops at Wakefield and Pontefract, to march to York and remain till reviewed by the Right Hon. Sir Charles Wills, and then return. Evidently there are some entries missing. The Right Hon. Sir Charles Wills was Lieutenant-General Wills, of Preston fame in 1715. The next change in the Regiment was this. Two troops from York to Doncaster, and one from York to Rotherham; the remaining three troops were left in their quarters, Pontefract (2) and Wakefield (1). This took place on 20 October. November 24 the Rotherham troop proceeded to Barnsley, and on 8 December the Pontefract troop marched to Chesterfield. The next entry, dated 10 January 1727, is as follows:

Kerr's Dragoons to march, viz.—
3 Troops from Doncaster and Barnsley to Stamford, 2 from Wakefield to Huntingdon and one from Chesterfield to Grantham.

February 21. One of the Stamford troops was moved to Melton and one from Huntingdon to St. Neots.

A war cloud had again arisen on the continent, and hostilities were expected to break out between the Emperor and Holland. In consequence the Regiment was augmented on 21 February to nine troops, and was ordered to hold itself in readiness to proceed to Flanders to assist the Dutch. A force of which Ker's Dragoons formed a part, and which consisted of four regiments of cavalry and eight of infantry, was prepared to embark for Holland, but the expedition was never despatched.

The warrant for the augmentation of the Regiment is dated January 31, and by it each of the new troops was ordered to consist of one quartermaster, two serjeants, two corporals, two drummers, one Hautbois and fifty private dragoons, besides officers. The existing six troops by the same warrant were also augmented from 45 to 50 private men each. Perhaps it should here be remarked that the hautboy was the instrument which in those days was used in the cavalry. Later it was superseded by the trumpet. The precise date when this change was actually made in the Regiment is not to be discovered in the records, but it was probably in 1767 or 1768. Cannon tells us that in 1766 'drummers on the establishment were directed to be replaced by trumpeters,' but he does not mention Hautboys.

On the death of George I, who it will be remembered died while on his journey to Hanover in the month of June 1727, his son George, Prince of Wales, succeeded to the throne as George II, In consequence the Princess of Wales, Wilhelmina Carolina, became queen. The style of the Regiment was in consequence changed, and from its former title of Her Royal Highness the Princess of Wales's Own Royal Regiment of Dragoons was in future to be known as The Queen's Own Regiment of Dragoons.

As has been already stated, the Regiment was ordered to be augmented in January of this year both in the number of troops and the strength of each. How soon this augmentation was carried into effect we do not know, but it had taken place by 14 October. Meanwhile the only changes of quarters or marches recorded are these:

26 June. The three troops at Peterborough were ordered to remove during the fair at that city, two to proceed to Uppingham and one to Kettering. On 3 Oct. the two troops then at Stamford were

ordered one to Biggleswade and the other to Woburn, both places being in Bedfordshire. On 14 Oct. however, we find that all the nine troops were in existence, as marching orders were issued as follows:

> The Queen's Own Royal Dragoons, commanded by Colonel Wm. Ker, to march from their present quarters:
>
> 2 troops from Woburn and Biggleswade to Kingstown.
>
> W troops from Peterborough, Huntingdon, St. Ives and St. Neots to Hounslow, Kingstown, the Brentfords and Richmond.
>
> 3 troops from Oundle, Uppingham and Kettering to Uxbridge, Twickenham, and Hampton and remain till Saturday the 28th October when they are to rendezvous upon Hounslow Heath to be reviewed by His Majesty, after which they are to return to their former quarters.

After the review, in which the Regiment acquitted itself so as to obtain commendation, three troops were sent from Kingstown (Kingston-on-Thames) on 1 Nov., 'on account of the fair there'; two troops went from Isleworth, Twickenham, Hampton, and Richmond, and four from Uxbridge, Brentford, and Hounslow. These various detachments proceeded to Wells (2), Blandford (2), Dorchester, Wimborne and Wincanton (1 each). At the same time a cornet and 20 men, with non-commissioned officers in proportion, received an order to march to Halesworth 'on the Revenue duty along the Sussex coast.' Halesworth, however happens to be in Suffolk. This does not account for the entire 8 troops as will be seen.

A paper in the Record Office dated 'Whitehall 4 day of Nov. 1727, gives a clue to what had happened. It is as follows:

> To all Magistrates, Justices of the Peace, Constables and other officers whom it may concern &c, &c.,' and it proceeds to give directions to 'impress wagons or carriages with horses, as shall be wanted to carry from Peterborough, Kettering, Oundle, St. Neots, Huntingdon and St. Ives, the Sick Men of the Queen's Own Royal Regiment of Dragoon. Commanded by Brigadier Kerr" to Dorchester, Blandford, Sherborne, Shaftesbury and Wimborne, the Officers paying for the same at the usual Rates.'

Evidently there had been an epidemic of some kind.

On 7 March 1727-8 the troop at Dorchester was moved to Sherborne and on 30 March from thence to Shepton-Mallet. On 9 May Her Royal Highness the Princess Amelia was about to pay a visit to Bath. Princess Amelia Sophia Eleanora was then quite young, having been born at Herrenhausen 10 Jan. 1710. She was the third child and second daughter of George II by Queen Caroline. For a long time she was designed to marry Frederick the Great, who corresponded regularly with her until his marriage in 1733. On her death in Oct. 1786, which took place at her house in Cavendish Square, it is stated that a miniature of Frederick was found upon her breast. When a princess went abroad in those days an escort of cavalry was ever in attendance. Hence the following orders to the Regiment:

4 April. A sufficient detachment of the Queen's Own Royal Dragoons, to receive from a detachment of Wade's Horse, at Newbury, Her Royal Highness the Princess Amelea, and escort her to Bath, and afterwards return to their quarters.

9 May. 24 dragoons from 3 troops at Wells to March under a Subaltern to Bath, there to remain during the stay of the Princes (sic) Amelia (correct this time) in order to attend upon her whenever she shall go abroad in her coach.

How long the first visit of the Princess continued does not appear, but seemingly the gaieties of Bath or its waters were to her taste, as we find her paying a second visit there in August of the same year.

17 August. A sufficient detachment from the troops at Wells is to receive from the Royal Horse Guards at Newbury, the Princess Amelia and escort her to Bath and then return. 24 dragoons from the troops at and near Wells to march to Bath to attend upon the Princess (as directed in the Order of 9th May last).

It was on the occasion of one of these visits that the Princess made a request to the notorious Beau Nash, the master of the ceremonies at Bath, the granting of which was, if not exactly refused, at any rate evaded by that worthy. Beau Nash, the so-called 'King of Bath,' ruled the festivities here with a rod of iron. One of his regulations was that on the stroke of eleven, the dancing a the pump room should instantly cease. One night the princess was enjoying her dance and begged the upstart autocrat to indulge her with one more.

Nash replied to Her Royal Highness that he hoped she would not insist upon it, because one deviation from the rules then established, would totally subvert his authority. How he had the impertinence to unfasten the white apron of the Duchess of Queensberry and throw it over among the waiting ladies'-maids is another instance of this person's good taste. White aprons, forsooth, were contrary to his regulations. The story of Princess Amelia and Nash is true, and reminds one strongly of the consequential pedagogue of old, who asked permission when in the presence of royalty to be excused from removing his headgear, lest his pupils should lose respect for him!

There are no other events to chronicle for the year 1728, except that on 16 May the detachment then on revenue duty at Halesworth was ordered to march to Wells and thence rejoin their troops, and that on 30 July the whole Regiment was concentrated at Salisbury for review. The review took place on 14 August, when the troops returned to their former quarters except that one of the Wells troops was sent to Glastonbury.

On 24 October the visit of the Princess to Bath terminated, when the detachment at that city was ordered to escort her as far as Newbury and then to proceed to rejoin their respective troops. On the day this order was issued (17 October) the three troops at Sherborne, Wincanton and Wimborne were sent to Salisbury, and the Glastonbury troop was despatched to Warminster.

The last event of the year was an order on 5 November for the troop at Blandford to proceed to Sherborne and one of the Wells troops to march to Shepton Mallet. An interesting petition from the Lord Bishop of Bath and Wells, dated from Westminster December 24 1729, is extant. It was sent by him at the request of the Mayor and Corporation of Wells and concerned the quartering of dragoons in that city. The Bishop states that when he was first consecrated to that diocese he found that dragoons had been quartered there for frequent and protracted periods. There were originally three troops, of which two weir removed and one remained. This remaining troop had been recently reduced and the inhabitants no longer had dragoons quartered on them. There was, however, a credible rumour that 'another is going to be sent,' and against their coming the inhabitants petitioned. There were, he states, two troops there in 1728.

Certainly in that year some troops of the Regiment were quartered at Wells,

The Bishop continues that several innkeepers had been 'forced to leave their business and shut up their houses.' The fact being that tho inhabitants would not frequent inns where soldiers were quartered and the billeting money was insufficient to recompense the innkeepers. He petitions for the city to be 'excused at any rate for a time from quartering any more' as they have been 'overburdened for several years past.'

We do not quote this paper with a view to offering any suggestion to the Chancellor of the Exchequer!!

1728-9. As the assizes were being held at Salisbury, according to the custom then prevalent, the three troops in quarters there at the time were ordered on 20 February to march to Shaftesbury (2) and Yeovil (1). The removal of troops from assize towns when the Courts were sitting there was usual and was professedly to avoid any suspicion of influencing justice. For similar reasons, soldiers were formerly confined to barracks, and territorials, even in these days, do not assemble for drill during an election, lest their presence in the streets should be supposed to affect the freedom of voters. Ordinarily the three troops thus dispersed would have returned to Salisbury on the conclusion of the assizes, but on this occasion the Shaftesbury troops were ordered to remain in their new quarters.

On 3 April, 11 April, 12 May and 6 June the following changes of single troops took place. Dorchester to Cerne, Warminster to Wincanton, Wincanton to Sturminster and Sturminster to Bruton.

On 14 August we read for the first time in the MS. Regimental Record that a troop had been at 'grass quarters.' This was the Bruton troop and was under the command of Captain Maxwell. It was ordered on that date to match to Wincanton, halt there one night, and then resume its grass quarters near Bruton. Probably the Regiment had been in grass quarters before, as was customary, but the dates have not been preserved, save in one instance.

On 26 September, 16 October and 15 November the Cerne troop marched to Wimborne, the Bruton to Warminster, the Shepton Mallet to Taunton and the Glastonbury to Shepton Mallet.

All this time a detachment of the Regiment had been employed since 14 April 1726 in revenue duty on the coasts of Hampshire and Dorsetshire. It was not until 20 November 1729 that these men were recalled and rejoined at Shaftesbury.

Prolonged absences like these can hardly have been conducive to the efficiency of the Regiment as a whole. Still it was the custom in those days to employ cavalry. In Ireland, when on the Irish Establishment, as some regiments were for at times forty years at a stretch, the outcome was much to deprecated. Instances have been known when a commanding officer never for years had an opportunity of seeing his regiment concentrated. But the reason for the recall of the revenue detachment was this.

It will be remembered that on 21 February 1727 the Regiment had been augmented from six troops to nine. It was now about to be reduced again to six. On 20 November a warrant was issued for the immediate reduction of the three youngest troops, and the remaining six troops were also to be reduced so as to consist of:

One captain, one lieutenant, one cornet, one quartermaster, two serjeants, two corporals, two drummers, one hautboy and 40 private dragoons.

It was long since the Regiment had been so weakened in strength. The troops reduced were those newly raised in 1727. The officers were relegated to half-pay as usual. The final move of the year took place on 23 December, when the Regiment was distributed in quarters at Blandford, Sherborne, Wimborne, Yeovil, Warminster and Devizes. These reductions were a great hardship and sometimes led to trouble. Two most interesting letters from General Wade on the subject of a mutiny in a regiment unnamed are in the Record Office and bear date December 3, 1729.

It appears that a solitary troop was quartered in a small town. Orders came for its disbandment. Without proper orders the officer had sold all the horses a few days before the date of disbandment. General Wade adds: 'I fear rather irregularly.' The troop when the day came was insolent and went so far as 'to insist upon terms inconsistent with His Majesty's Orders of Reduction;' and refused to be drafted into other troops.

General Wade was sent to deal with the matter. He ordered up a squadron of old troops under the major then in command of the regiment. This clearly shows that it was a newly raised troop which was to be disbanded. Orders to arrest the ringleaders were given, and the men were put in guard. But the town was too small to put up a squadron and a disbanded troop as well, so he moved all to a larger town.

The major who came in command of the squadron to arrest the mutineers actually read the orders of reduction on his arrival, at the head of the regiment and in presence of a Justice of Peace. The troop was disarmed, but still showed great insolence to the officers, who did not, says General Wade, behave with the 'prudence and precaution' they should. Clearly the mutinous troop should have been sent to headquarters. Two courses were open to Wade—he must either try the men by court-martial, or adopt the less rigorous procedure in the civil courts. General Wade decided on the latter course, for, as he says, he had grave doubts whether the mutinous troop was not already disbanded, and if disbanded no court-martial would have authority to deal with the case. Neither could he confirm a death sentence or other punishment on men not under martial law.

The prisoners soon repented of their behaviour, and sent a message to Wade, begging to see him and asking pardon. They also assured him that they were ready to comply with His Majesty's order. Wade accepted their expressions of regret, for, as he says, 'their conduct had not been very outrageous as far as acts of mutiny,' though 'very seditious and insolent to their officers.'

Next morning he drew out the troop, and in person read out the order disbanding them again. They behaved, he states, in a 'Decent and respectful manner and continued so.' General Wade promises to come to London shortly and to make a full verbal report then.

It will be remembered that certain dragoon regiments were in February 1727 augmented from six to nine troops, and that in 1729 they were reduced again from nine troops to six. As a light cast on the military administration of those days, the insertion of this narrative is not without value.

The Regiment then remained in quarters till 21 March, when the following changes took place—Devizes to Shaftesbury and Warminster to Bruton. Four days later the Wimborne troop marched to Cerne. On 22 September the Bruton and Cerne troops proceeded to Wells, the Yeovil troop to Shepton Mallet and the Sherborne troop to Warminster. Lastly, the Shaftesbury troop marched to Wincanton on 28 November.

1731. On 31 March a detachment equal in strength to a troop was ordered to march to Lewes in Sussex to be employed on revenue duty. April 15. The two troops at Wells, the troop at Shepton Mallet and the troop at Wincanton were ordered to Salisbury.

On 1 May the whole Regiment was concentrated at Salisbury and received orders to proceed, three troops to Uxbridge and the other three to Colnbrooke (Bucks.) and Longford (Langley?, Bucks.). The Regiment was to rendezvous on Hounslow Heath, there to be reviewed by the King on Saturday, 15 May After the review, in which the Regiment acquitted itself well, the troops returned to their several quarters in the West of England, Wells (2), Blandford, Yeovil, Sherborne and Somerton (1 each).

At the review the strength of the Regiment was only 200 men. Forty more men were now to be taken off on revenue duty. Assuredly the owlers and smugglers must have been particularly active around Lewes, as on 26 May 'another detachment equal to that ordered on 31 March last' was dispatched to Lewes on preventive duty. From a military point of view this was truly a most deplorable state of affairs.

On 10 June the Blandford troop was moved to Crookhorne (Crewkerne, Somerset). On 2 Jury an order was issued which was subsequently cancelled by which it was intended to remove the whole Regiment to the neighbourhood of Lewes, where practically a squadron was already quartered. The places selected were Lewes, Hailshom, Eastbourne, Battle (near Hastings) and the villages adjacent.

Probably the common sense involved in concentrating the Regiment to a certain extent in one district was too much for the authorities, and in lieu of proceeding to Sussex the Regiment remained divided between Shepton Mallet. Glastonbury and Warminster, the order being cancelled.

At length in 1732 the Regiment was removed from this district and went into quarters on 24 March at Cheltenham, Leominster, Hereford and Gloucester. The Cheltenham troop was, however, shifted on 13 May to Moreton (Moreton-in-Marsh? Gloucester, or Moreton-on-Lugg? Hereford, but not Moreton, Dorset), and then on 24 June was moved on to Tewkesbury.

The Regiment had now been for ten years in England and, as

will have been noted, had during that period had a most extensive experience of marching hither and thither almost perpetually, with intervals only of sometimes a month or two, at others hardly a week or more. But these were days when barrack accommodation in England was scanty and barrack accommodation for cavalry specially so. Hence to billet troops became a necessity. Now the billeting system was unpopular with those upon whom troops chanced to be billeted. But with a view to preventing more hardship than need be to the civil population this method of changing of quarters perpetually was adopted. At any rate, it prevented too large a proportion of complaints, petitions, &c.

As a system it was a bad one, but at that time no other course except that of building sufficient barrack accommodation was open to the authorities—and, as usual, money was scarce.

The petition of the Bishop of Wells already noticed throws a good deal of light on this phase of military life in those days.

On 17 April 1733 the regiment received orders to march forthwith to North Britain *via* Berwick, at which place on arrival they were to find orders from Lieutenant-General Wade as to their further procedure. The order was, however, somewhat varied, for on the arrival of the first three troops at Berwick, they were despatched to Dunbar and Haddington, while the other three, having marched to Wooler (close by the battlefield of Hedgeley Moor), were directed to change their route and proceed via Coldstream to Dunse. The ford at Coldstream was one of the most favoured crossing places into England of both regular invading Scottish armies and countless bodies of raiding freebooters of old.

The Lewes detachment left that district to rejoin in Scotland, 15 May 1733. In Scotland the Regiment remained until 10 April 1735, when it again returned to England. Berwick was reached on this date, and thence the Queen's Own Royal Regiment of Dragoons proceeded south, two troops being quartered at Leicester, two at Loughborough, one at Ashby-de-la-Zouch and one at Burton-on-Trent.

On 3 May the Burton troop was ordered to Leicester to join the two troops already there, and on 13 May the Loughborough troop was sent to Lutterworth. Revenue duty again fell to the lot of a detachment, which in strength equal to a troop was ordered to Boston, Lincolnshire, for that purpose. This order was dated 26 June, and on the same day a larger detachment consisting of '90 privates with commissioned and non-commissioned officers in proportion' was ordered to march to Norwich 'on the Revenue duty on the coast of Norfolk. The Regiment was thus deprived pro tem. of no less than 130 officers and men.

The only other move during 1735 took place on 6 Nov., when the the Ashby-de la-Zouch troop proceeded to Leicester.

The events for 1736 are few.

On 27 April the revenue detachment for Norwich was withdrawn and the men rejoined their various troops. An entry dated 3 May gives a reason for their return. On that day the whole Regiment, that is to say all except the Boston detachment, was ordered to assemble at Leicester, there to be reviewed by their Colonel, now Major-General the Hon. William Ker. After the review the various troops returned to their several quarters.

Three more entries only concern this year. On 13 May one of the Leicester troops moved to Melton Mowbray. On 15 July as the assizes were being held in Leicester the troops there were withdrawn, but their destination is not stated. Lastly, on 27 Sept. the Loughborough troop was ordered to proceed to Ashby-de-la-Zouch.

1737. On 14 April the whole Regiment was employed on revenue duty; three troops being sent to Norwich and the other three to Lynn, Bungay, Beccles and North Walsham for that purpose. On 9 June a detachment of 50 dragoons with commissioned and non-commissioned officers in proportion was formed and marched into Lincolnshire on the same duty. This detachment was thus employed till 13 October, when it received orders to return to Norwich.

1738. The Regiment was now ordered to march from Norfolk into Herefordshire and Gloucestershire. On 20 April, therefore, it proceeded to Hereford (2 troops), Gloucester (2), Ross and Tewkesbury (1 each). Apparently they were permitted to remain in these quarters for ten months, an almost unexampled period of time for many years.

1739. Feb. 22. The Ross troop was moved to Leominster. April 3. A detachment equal to a troop was sent to Lewes on revenue duty. 22 Sept. The Leominster troop was sent to Hereford and the Tewkesbury troop to Gloucester.

Dated from Kensington, 5 Sept. 1739. We have an Assignment Warrant for the colonels of 8 regiments of dragoons in Great Britain, the Colonel of the Queen's Own Royal Regiment of Dragoons among them. This warrant gave the colonels permission to make assignments of their off-reckonings for the new men from 25 June 1739 to 24 June 1741.

An additional serjeant and 10 men per troop had recently been raised, and though assignments had been made for the old men in the Regiment no provision to defray the expense of clothing these new men existed. This warrant supplied the need. It was issued after a request from the colonels of the respective regiments and was considered by the King to be just and reasonable. The clothing was directed to be prepared forthwith and ready to be inspected by the Board of General Officers.

1740. On 1 Oct. the Queen's Own Royal Regiment of Dragoons, commanded by Major-General Kerr, was ordered to march from the camp near Kingsclear (Kingsclere, Newbury, Berks.) and proceed to Cirencester (2 troops) and Tewkesbury (1), Monmouth (1), Farringdon (1), Fairford (2) and Ledlade (½). Three days later the destined Monmouth troop was sent to Highworth instead. This is the first entry recorded in the Record for the year 1740. It may be, however, well to state that considerable exertions for those days had been expended in the formation of camps during the preceding few months.

As early as 15 March three were ordered to be held in Hyde Park, Hounslow and at Blackheath. Later, three more were pitched, at Newbury, Windsor and another at Hounslow. The Regiment was sent to the Newbury camp, which was actually formed at Kingsclere close by. There four other regiments of cavalry were collected and four infantry regiments as well. Lieut.-General Wade was in supreme command.

Two other changes of quarters took place during this year, one of the Cirencester troops being removed to Burford on 11 Nov. and on 9 Dec. the Highworth troop marched to Bromsgrove.

17 Dec. 1740. The Tewkesbury troop was ordered to march immediately to Evesham 'to assist in putting down the riots in that neighbourhood.' These were corn riots, which had been prevalent in several places during the year. The worst were at Newcastle-on-

107

Tyne 7 to 25 June, where, owing to the militia which had been called out being unwisely disbanded, the mob broke windows, burnt public records and looted £1800. Some gentlemen armed themselves and fired on the rioters. At Wisbech in the Isle of Ely similar scenes took place. Corn was seizes and sold for 1d. to 4d. per bushel by the rioters. Next they levied £200 on the town. Eventually 500 men were raised who secured 60 of the disorderly persons. At Norwich wheat was at 16s. a comb. Dragoons were called out and some rioters were arrested; the mob then broke open the gaol and released them. The dragoons fired m the mob, killing three men, two women and a boy, many others being dangerously wounded. There were also riots at Derby, Northampton, Wellingborough and at Evesham as before stated.

1741. 7 Feb. 25 Dragoons of the Burford troops were ordered to march to Chipping Norton; and the Cirencester troop on the same day to proceed to Tewkesbury. 19 Feb. 20 Dragoons of the Farringdon troop were sent to Highworth. 5 March. The troop which was divided between Fairford and Lechlade marched to Tetbury and Hampton. 4 April. The Tewkesbury troop was moved to Stroud and Painswick. 11 April. The Farringdon and Highworth troop was to be thus divided, 20 men to Witney and the remainder to Woodstock. On the same date the men at Evesham were withdrawn, 15 being lent to Pershore and 15 to Alcester, from which it would appear that the troop was ten below strength. 14 April. The troop at Burford and Chipping Norton march to Stow and Moreton; and the troop at Stroud and Painswick to Campden (Gloucester) and Winchcombe. Also 30 men were dispatched from Tetbury to Dursley and Woolen. These movements of small bodies in different directions were evidently measures of precaution against rioting.

On 18 April 1741 the Regiment suddenly received orders to march from their present quarters to North Britain. They started forthwith, but having proceeded for some days in the required direction were with equal suddenness halted and ordered 'to be disposed of in the several towns and villages most contiguous to the place where this shall reach.' Two days later they were all sent back again to their former quarters.

Apparently the halt was called at Nottingham, Leicester, Chesterfield. Lutterworth, Loughborough and Derby, as from those places they are ordered to Moreton, &c, Tetbury, &c, Bromsgrove, &c, Woodstock, &c, Campden, &c, and Evesham, &c.

On 2 June the Regiment was again ordered to march to North Britain, and again it started. On 16 June it was again halted and again sent back to its old quarters—'the respective places from whence they began their march.' By 27 June they had returned as ordered, and the Woodstock troop was then moved to Highworth On 12 August 1741 Major-General Sir John Cope, K. B., succeeded to the Colonelcy of the Regiment owing to the death of Lieut.-General the Hon. William Ker. For details of Lieut.-General the Hon. William Ker's life and career see Appendix 1. 19 Sept. A detachment consisting of 1 officer, 1 serjeant and 20 dragoons was ordered to assemble at Tetbury and march to Deal on revenue duty. 23 Oct. Twenty men of that part of the troops at Campden and the same number of that part of the troop at Highworth were sent to Skipton-upon-Stour and Witney respectively.

For the third time this year, by an order dated 7 Nov., the Regiment was marched off to Berwick en route for North Britain; it being ordered that the several divisions should halt 'as long as necessary to avoid the floods or other accidents on their march to Berwick. In North Britain the Regiment at length arrived, though the date is not known. However, it is recorded that on 27 April 1742 'The Queens Royal Regiment of Dragoons' on their arrival back in Berwick were to march forthwith, three troops to Colchester and one each to Chelmsford, Ingatestone and Burntwood, and Bocking and Braintree. Meanwhile a detachment of the Regiment had somehow been posted at Ashford without any clue as to which Ashford is meant, probably in Kent.

However on 6 May these men were ordered to march to Ipswich. On 1 July one of the Colchester troops was ordered to Witham. War was now again in progress; and on 6 August the Regiment was once more selected for active service abroad and was under orders to hold itself in readiness to embark for Flanders.

Note :—When we consider the multitude of provincial towns in the United Kingdom at which the Regiment was stationed during this period of its existence, that is to say between 1716 and 1742, it is impossible to avoid being struck by the fact that even in these days very many of them are places of no little architectural beauty and antiquarian interest, not to mention their historical and military associations. How much more beautiful and interesting they must have been in the first half of the eighteenth century it is not easy to gauge. In not a few of these places relics of buildings are to be found even yet which cannot be found elsewhere. Take, for instance, the Roman gate of Colchester, now overshadowed by the hideous Jumbo water-tower; Woodbridge with its quaint overhanging weigh-bridge: it was in position but a few years since, and in use. Orford with its quaint polygonal castle and sadly dilapidated church; and was not the famous man-fish caught in mediaeval times in the North Sea off Orford Ness? Stamford, a small town crammed even yet with remains of the England of the past, and Grantham but little behind it in interest; Hexham, one of the most attractive of the northern towns; Northallerton with its church porch stones, scored as are some on Sedgemoor, with the marks where weapons have been sharpened thereon; Wakefield with its memories of the Yorkist defeat outside the mighty ruins of its once mighty castle, and with its chapel adorned bridge: this last, one of the three chapelled bridges which now remain. A few years later the Regiment was at St. Ives in Huntingdonshire, where another chapel is to be found on the bridge, though in the eighteenth century this had been raised in height and converted into a lighthouse. Newbury with its quaint old hall, with its memories of two battles, and with Donnington Castle hard by at Speenham Land; Knaresborough, picturesque and still guarded by its noble though slighted castle, and as yet unknown to fame as the habitat of Eugene Aram; Pontefract, the Pomfret of old, and one of the strategic points which in the Wars of the Roses the contending parties were ever striving to seize. These are but a few of the places at which the Regiment was quartered. And be it observed we have said nothing of cathedral cities, neither have we mentioned Tewkesbury, Evesham, Taunton, Berwick, Glastonbury, Sherborne or Devizes and others, all of which have their architectural features as well as their military memories.

Dettingen
1743

In the preceding chapter we have told how the Regiment, whose colonel was now Major-General Sir John Cope, left Scotland and, marching south, was shortly afterwards selected to form a part of an army of 16,000 men destined for active service in the Netherlands under the supreme command of Field-Marshal the Earl of Stair. The object of this expedition was to support the interests of the Queen of Hungary and Bohemia then threatened by France and Bavaria. The design of this powerful combination was to deprive the Archduchess Maria Theresa of the Kingdoms of Hungary and Bohemia.

The quarters of the Regiment on 27 April 1742 were as follows: Three troops at Colchester, one at Chelmsford, one divided between Ingatestone and Brentwood, and the last divided between Booking and Braintree, At the time a detachment of the Regiment was also at Ashford in Kent, probably on preventive duty, and this on 6 May was ordered to Ipswich. On 1 July one of the Colchester troops was sent to Witham, and thus distributed the Regiment remained until 6 August, when the route arrived.

On that date the several troops were ordered to march from their present quarters as follows, and there to remain until the date of embarkation was received. Four troops to proceed to Rochester, Strood and Chatham, and the remaining two troops to Gravesend and North Fleet.

In a few days the Regiment duly embarked, its destination being Ostend. On arrival the Regiment marched a few stages up the country and was then halted. No active measures were undertaken during the remainder of the year, as according to the military notions of those days it was too late, and so going into winter quarters the force remained stationary and inert until February 1743.

The political position, which can, however, be but briefly sketched here, was somewhat curious. The Emperor Charles VI had died in 1740, leaving Maria Theresa, his daughter, the sole heiress of his dominions, and as such she had been recognised by the Powers of Europe, a recognition guaranteed by the Pragmatic Sanction. The Elector of Bavaria, however, took a hostile view of the situation as regards Hungary. France hating Austria was prepared to support the Elector.

The various minor Genoa States considered this an opportunity for possible aggrandisement, that is to say, they hoped for a chance of sharing the spoils, if any, should any grave political changes occur. The King of Prussia began overt acts of war by invading Silesia, where he overthrew Austria at Mollwitz in April 1741. It was the lighted torch which set Europe ablaze, though the conflagration took some time to develop dangerous proportions.

France, who not only had designs against Germany but also hoped to strike a blow at England through Hanover, intrigued all round in an endeavour to get the Pragmatic Sanction abrogated, and also to obtain general European support for the candidature of the Elector of Bavaria (Charles Albert) as against the Grand Duke Francis of Lorraine, the husband of Maria Theresa, who claimed, and rightly, the Imperial Crown. But the point of the whole intrigue, though diplomatically concealed, was a design to divide Austria between Prussia and Saxony and to divide the entire Empire into four kingdoms—Bavaria, Hungary. Saxony and Prussia, each of which would be too weak to lead and too jealous of each other to unite against any separate or common foe. Hence France would become the predominant partner in the whole of that portion of Europe. Bavaria, Saxony. Prussia and Spain acquiesced in this charming arrangement, but England, backing as it would Hanover, adhered to the Pragmatic Sanction. Intrigues on both sides were now the order of the day. England's attempt to detach Prussia from this unholy alliance failed. King George then went over to Hanover to gather forces to support Maria Theresa.

France took the field, and in two directions—one army crossed the Rhine to join Bavaria and march on Vienna, the other marched directly on Hanover. This action, unforeseen as it was, took King George by surprise, and he was compelled to temporise.

A year's neutrality for Hanover was agreed upon, during which England engaged neither actively to assist Maria Theresa nor to vote for her husband. It was an arrangement which was equally despised and condemned in both England and Austria. Walpole then went out of office. Public opinion in England was strongly in favour of Maria Theresa, and Parliament voted her a large subsidy which amounted to no less than £500,000. This pecuniary assistance was later increased by the despatch to Flanders of the 16,000 men previously mentioned, as a force of so-called auxiliaries, and a vote of £5,000,000 to prosecute the war, or rather a war, for at the time we were technically at peace with France; and our force when it reached Flanders was on paper supposed to be an auxiliary army despatched thither to lend moral rather than active support to Maria Theresa. Lord Stair was appointed to the supreme military command of this force, but he was also entrusted with political duties. His first diplomatic employment was to persuade or endeavour to persuade the Dutch to permit the occupation of Nieuport and Ostend by the English force, bases of operations against the French being needful, seeing that the Austrian Netherlands were in danger.

For some time his efforts were fruitless; but on the news of the Austrian defeat by Frederick of Prussia at Chotusitz a grudging consent was obtained from the States-General. The situation was very serious The French, under Marshal Maillebois, to the number of 40,000 men, lay between Hanover and the coast, should any British force be landed, and the preparations for war then in progress in Britain could point to no other design but a landing in the Netherlands.

As we have said, the force sailed from England and duly disembarked, but it is to be observed that it was self-interest alone that prompted the Dutch to acquiesce in the arrangement. For the Pragmatic Sanction they cared nothing, but of a French invasion of Holland they had a holy terror.

Meanwhile, the French had not had it all their own way in

Austria, more than meeting their match in the Austrian General Khevenhüller. In Bohemia they had met with more success, having captured Prague, where a force under Marshal Broglie was established, but Broglie was not by any means a great soldier, and his occupation there caused but little anxiety.

The hostility of Frederick of Prussia was another matter, and of a far more serious nature; and while this hostility continued the serious nature of the crisis remained. Vainly did Stair endeavour to detach the warrior king; and vainly did he, with the use of every diplomatic wile, endeavour to persuade Maria Theresa to come to terms with Prussia. But that really remarkable woman would have none of it, till after her crushing defeat at Chotusitz. Then by the surrender of Silesia to Frederick, on the conclusion of the Treaty of Breslau, his active opposition was removed.

It was now possible for the entire Austrian force to be directed against Broglie, and it looked as if he be could be annihilated. In this case Austria could have invaded France via the Palatinate and along the Moselle, while England and Holland could either join them or march as invaders on Paris from the north on their own account. Stair advocated such a proceeding, and had his view been adopted there is little or no doubt that France would have been crushed.

As it was, Austria speedily tackled Broglie, and Maillebois was compelled to hurry to his assistance, thus to all intents and purposes leaving the path to Paris open, for the French garrisons on the frontier, save at Dunkirk, were too weak to offer any real opposition. Moreover these garrisons were scattered and could have been dealt with in detail. Dunkirk was, however, a very strong place. Its fortifications had of late been greatly increased and strengthened; but this had its drawbacks, for it now needed a garrison of 40,000 men to hold it.

Dunkirk, then, Stair begged, prayed and implored the British Government to besiege, and at once. He was right, for the French must then either lock up this huge garrison in the town or withdraw it. King George, however, thought differently; his view was that the fiction of being at peace with France must be maintained, and to besiege Dunkirk, a French place of arms, would certainly be an act of war. Delay after delay took place. The veteran General Wade, to whom Stair's plan was referred, threw cold water on it, in his ever over-cautious way. The season of 1742 passed by, nothing was done, the troops already in Flanders went into winter quarters till the following year, and Stair had perforce to possess his soul in patience.

Meanwhile the French had not been idle, and a large force had moved to the Moselle to protect Lorraine; another, collected from Flanders, lay on the Meuse prepared either to join in the same object or to carry the war across the River Neckar, in the somewhat improbable event of an invasion by the Allies over the Rhine south of Cologne.

Stair then proposed a march to the Danube—again he was thwarted. Finally, by the orders of King George, he occupied the high ground of Mainz, which commanded the junction of the Rivers Main and Rhine. Gradually, and by slow degrees, the armies of England, Hanover, and Austria effected a junction on the north bank of the River Main. The French, under Noailles, were in position near Spires on the Upper Rhine, and numbered 70,000 men; and they were in the hope of giving succour to Broglie could they but join hands with him. Stair was for an immediate advance. He wished to cross the river to follow its course downwards, and to establish himself between Oppenheim and Mainz. This would effectually have held Noailles in check, and more than that would have protected Bavaria and have enabled him to attack the enemy at will. A crossing of the Rhine would have been feasible, and having been accomplished would have opened the road for an invasion of Lorraine.

On 3 June 1743 the Allies actually began to cross the Main. But Noailles had not been idle, and had already set his force in motion, and instead of crossing the Neckar had passed through Darmstadt towards Frankfort intent on attacking the Allies. Stair proceeded to await him at a spot where the road debouched from a forest. The Austrian commander, whose military genius was not excessive, here, either through timidity or excess of caution, would by no means co-operate energetically in the movement. He considered, nay, made up his mind fully, that Stair was bound to be defeated, and in consequence withdrew his cavalry across the river, leaving only his infantry to engage in the expected struggle. Noailles marched on till he came in view of the Allied Army, and then, not appreciating the prospect, withdrew without even offering battle.

Stair's position was hardly to be envied. His grasp of the situation was complete and his proposals were sound, but King George, who had by this time arrived in Hanover intending to take command in person, thought differently. His various military advisers, both British and foreign, obsequiously echoed his opinions, and Stair was hampered in every way whenever he suggested any course of action. His very crossing of the Main, by the way, was in direct disregard of orders, for he suppressed the order forbidding him so to do; but his doing so was justified in the event. He, however, re-crossed the river when Noailles declined battle. At this juncture King George arrived, and what is more took over command.

The Allies were then encamped in a most unfavourable spot, a plain between the River Main and the Spessart Hills, hills thickly covered with wood and near to a place called Aschaffenburg.

This was not well in itself, but what was worse was that supplies were very short, and on this, the right bank of the Main, forage was not to be obtained. At Aschaffenburg there was a bridge, and this on the side of the Allies was fortified. Forage could be obtained across the river, and the battery thus erected would protect the passage.

Starvation speedily stared the Allied Army in the face. Their magazines were at Hanau, a town on the same bank of the river lower down, and distant by road some thirteen miles. Half-way on the other bank of the river was Seligenstadt, a place which was seized by Noailles and near which, lower down, he had thrown two bridges across the Main. It was possible, therefore, for him to cross at will and cut off the Allies should they make for Hanau by the road. Four miles from Aschaffenburg along the road is a village called Klein Ostheim, to which for a portion of the way there is a loop in the road, which at one part passes quite close to the river.

Three miles further on is Dettingen, which is distant as the crow flies about two and a half miles from Seligenstadt. Hanau lies at between five and six miles beyond Dettingen. The country is very woody on both sides of the river, and the river itself winds considerably.

Noailles pitched his camp on some comparatively level ground facing Aschaffenburg, where the Main makes a long narrow loop. From this spot he proceeded to make such dispositions as would in his opinion render defeat to the Allies an absolute certainty directly they began to move towards Hanau; and that, being starved out, move or try to move thither they must Noailles first erected a battery on his side of the Aschaffenburg Badge, thus blocking it. He also posted sundry guns at intervals along the river on his side, so disposed as to rake the road along which the Allies must march. Some accounts state that he so employed five batteries, others four. By means of his two bridges below Seligenstadt he prepared to send across the river a force of 28,000 men, and these the pick of his troops, under Count Grammont to block the road hard by the village of Dettingen.

Noailles had plenty of time to effect these preparations, as for a week the Allied Army lay in Aschaffenburg lacking provisions, with discipline falling to pieces and with the troops in their need resorting to attempts at at pillage, though there was little to be obtained thereby. Something had to be done. Affairs were in a desperate state by 26 June, and that night an immediate retreat to Hanau was determined on. Early on the morning of the 27th the retreat began, and already the signs of motion in the Allied camp had been reported to Noailles. He immediately in person delivered his orders to Grammont to cross the river, and assigned to him a spot close to Dettingen where he could wait for and attack the Allies as they crossed a little bridge, for here the ground was boggy and the road to Hanau led over a small brook by a bridge, at this place the only crossing. Truly according to all precedents the Allied Army was in a very tight place.

Ahead was Grammont with 28,000 men; on one side the Main whose banks were lined by the guns aforesaid, on the other side steep woody hills, and in the rear Noailles prepared to cross at Aschaffenburg with 20,000 troops of the French Army as soon as the Allies had cleared out of that delectable spot.

When the retreat to Hanau began the British cavalry led the van, followed by the Austrian cavalry. Next the British infantry and the Austrian infantry. Lastly the British Guards and the Hanoverian infantry and Hanoverian cavalry.

117

No sooner were the Allies well on their way than Noailles advanced on Aschaffenburg as predetermined. It occupied about three hours for the Allies to reach Klein Ostheim, where about 7 a.m. the retreating army had to make its passage by a narrow road.

When the cavalry in the van had passed through the place it was halted to permit the slower passage of the infantry. With this object, therefore, and to clear the road, it was drawn up between the road and the river and facing the latter. The Allies once through the village were now within the range of the guns so skilfully placed by Noailles, and the cannonade began. Not only this, too, but some guns in the rear also opened on them. These hitherto had been unable to come into action as the line of march of the French on Aschaffenburg masked them.

Of the fire of these rear guns the Allied baggage got the chief benefit, and much confusion ensued.

For more than an hour the retreating Allies were being industriously pounded while endeavouring to form in line of battle in a spot so narrow and contracted that any formation was almost an impossibility. At length the British artillery, which was not at first available, was hurried up to the front, and after a duel succeeded in putting the French guns out of action. All through this time of stress the baggage wagons and animals were being gradually withdrawn and were finally massed together on the other side of the road in a wood which, having boggy ground on either flank, seemed to promise security for, at any rate, a brief time. Moreover, too, it was out of the range of cannon shot. King George, whose personal exertions on the occasion were very considerable, managed at length to get his army drawn up in something like proper formation, but it was formation of by no means the regular kind. The troops were stationed as follows:

Left front line: 33rd Foot; 21st Foot; 23rd Foot; 12th Foot; 11th Foot; 8th Foot, and 13th Foot. On the right of these an Austrian Brigade of infantry, succeeded by the Horse Guards, the Life Guards, the 6th Dragoons and the Royal Dragoons. As a second line, and immediately behind the extreme left, came in succession the 20th Foot, 32nd Foot, 37th Foot, 31st Foot, and the 3rd Foot.

The second line of cavalry consisted of the 7th Dragoon Guards, the King's Dragoon Guards, the 4th Dragoons, 7th Dragoons and the Royal Scots Greys. It is to be noted that the extreme left of the British first line did not rest on the river, but left a gap of from 200 to 250 yards unoccupied. In front of the Allies the French under Grammont were drawn up also in two lines but with a reserve in addition in the rear. Their infantry was in each case in the centre and their cavalry on each flank, the extreme right being occupied by the cavalry of the Maison du Roi.

We must now return to Grammont, whose orders were to await the Allies at the bridge. This he had proceeded to do, but wearying of standing, as he supposed, idle he had in disobedience to his orders crossed the bridge and advanced right through the village of Dettingen, and was at the moment of the formation in line of battle of the Allied Armies endeavouring to get his men also into some semblance of order. This disobedience of Grammont's caused or mainly caused the failure of the well-matured scheme of Noailles. King George awaited no attack, but attacked himself, and took every advantage of the fact that discipline in the French Army was slack, the officers incompetent and regiments and brigades in considerable confusion. Indeed it is stated that the French Household Cavalry never really took its allotted station but marched and counter-marched in a meaningless way in front of the French infantry and between the lines of both armies. This they could safely do, as the lines being drawn so near together the French guns across the river were perforce silent for the moment.

The position was now this on the Allied left: opposed to the British infantry were huge masses of cavalry, and a gap between the extreme left and the river of some distance. General Clayton, who was in command, perceived this and despatched a request for cavalry to close this interval. The 3rd Dragoons were sent to him, and they and the 31st Foot had for some time to experience the effects of the destructive fire from the French guns which had now again got into action. But a small and weakened regiment of cavalry had a hard task ahead, which was nothing less than to cope with their numerous, fresh and heavily armed opponents.

The advance of the Allies was resumed. King George full of ardour was well ahead of the line. Without orders some of the infantry opened fire, and this had the effect of frightening the white charger ridden by the King, which incontinently turned tail and bolted to the rear, despite all the efforts of its royal mount. It was, however, caught, and the King dismounting fought, and fought bravely, for the rest of the day on foot. Again the Allied line halted, loaded, redressed and again advanced. And now the French put themselves in motion on the right centre where the Infantry of the Guard were posted. They opened fire but did little damage. The British replied, not in a fitful or disorderly way, but with volleys regularly and precisely delivered; reloading swiftly and displaying the utmost steadiness. Some cheering now took place, but this was checked, until by the signal, it is said, of Stair himself, who waved his hat, a simultaneous cheer burst out all along the line. Whether it was the effect of the volleys or that of the cheering who shall say, but the French Infantry of the Guard quailed and fell back to reform behind their cavalry.

The British still advanced, but another, anew and potent danger, was in front of them. Down on them came the cavalry of the Maison du Roi; and to meet them three regiments of infantry, one the 33rd which had already lost many men, and two squadrons, not by any means strong squadrons either, of the 3rd Dragoons. General Clayton at once discovered the peril in which his men stood, for he could not only be attacked in front but could be outflanked owing to his still improperly secured left.

But, notwithstanding, he ordered the 33rd and 21st and the 23rd and the 3rd Dragoons to face the music, and face it they did right nobly. Still, knowing that his extreme left was really unprotected and that it was open to a flank attack Carton despatched urgent requests for a reinforcement of cavalry to secure it. This reinforcement at length came, but before it arrived much had happened. Down on the British infantry and the two squadrons of the 3rd Dragoons dashed the masses of the French cavalry. Out to meet them rode the 3rd Dragoons, and the odds were two squadrons of British cavalry that had been pounded by artillery, against nine squadrons of their enemy not only fresh but ate clad in breastplates. The opposing forces met and at a brisk pace.

The 3rd Dragoons, despite all disparity, cut their way through their opponents, but not without heavy loss. It was then the turn of the 33rd, who had to withstand the onset of their mounted and protected foes. On rode the French, but were met with volley after volley—and down went horse and man. Another body of French cavalry took on the 21st and 23rd, firing on them with their pistols first and then in accordance with time-honoured custom flinging the empty weapons in the faces of the men prior to an attack on the ranks with the sword. It may be remarked that this custom was in vogue in England as far back as the days of the Great Rebellion. Again the French cavalry were met by a steady fire, and the fire prevailed, for they fell back somewhat in disorder. By this time other forces were at work to the detriment of the British. A French infantry regiment opened a flanking fire on the 3rd Dragoons and many fell. It looked as if the British left could be outflanked, and had this happened nothing but disaster could possibly have supervened.

The 3rd Dragoons were by this time in such a condition from casualties that further active participation could hardly be expected; yet, rallying their shattered force, they not only charged the French cavalry once again but twice, and cut their way through.

But now of their officers only two remained, and more than half their men had either been killed or wounded. Two of their guidons had been shot or cut to pieces, and a third had fallen from the hand of the officer who carried it, his hand having been slashed with a French sabre. It was then that a trooper in the Regiment, by name Thomas Brown. observing the guidon lying on the ground, dismounted to regain it. A Frenchman cut at him and severed two of his fingers; still he managed to regain his horse, but the animal bolted headlong into the French and through them to the rear of their ranks.

A Frenchman picked up the guidon and rode off with it. Brown saw him, and though wounded went in pursuit. He killed the man, and managing to grip the guidon he thrust it between his thigh and the saddle; then, alone and wounded as he was, by a desperate effort he cut his way back to the remnant of his corps. He saved the guidon, but in so doing received more than half a dozen wounds. Trooper Brown recovered, and after some weeks was sent home, where a post was found for him in the Household Brigade. But the brave fellow did not live long to enjoy it, for he died the following year.

It has been told how reinforcements of cavalry had been applied for by General Clayton, and these were sent from the right of the British line. Headlong at a gallop the 1st and 7th Dragoons came up and dashed into the French cavalry, but the charge was not successful, probably because it was not delivered in regular line, and also it must be remembered that the breastplates and helmets of the French were no small protection both against sword and pistol. Shock tactics have their value, but they must be applied correctly to be effective. For a space the two regiments were repulsed and retired to rally. Following on them came the British Horse Guards, who shared a similar fate and from the same reasons.

Believing that they had thus disposed of the British cavalry, the Frenchmen now advanced for a second attack on the 21st and 23rd Foot and this time with partial success. It was, however, but momentary, for though the cavalry had cut through their lines the British infantry stood firm, rallied, faced inwards, and having thus enclosed their adversaries shot them down without mercy. More British cavalry now came up, the 4th and 6th Dragoons, and also two regiments of Austrian cavalry. Twice these attacked the French and twice they could not prevail. A third attack now took place, in which the rallied squadrons of the 3rd, the Horse Guards, the 1st and the 7th joined. This was a last and supreme effort and the British finally were victorious.

It was altogether a curious battle, for on the right of the British line a French attack had been repelled with ease. The enemy indeed, did not appear to care seriously to face the volleys of the British infantry. A strange episode, however, took place, and it was this.

Dashing between the opposing lines of infantry, and en route receiving the fire of both friend and foe, the French Black Musketeers charged from what was their station on the right of the French line and hurled their squadrons on the Royal Dragoons who were posted on the extreme British right. It was a mad enterprise at best. The chance now offered to the Allies was at once perceived by Marshal Neipperg the Austrian Commander. He ordered the British cavalry to make a frontal attack on the advancing Black Musketeers, while his own threw themselves on the flank of the Frenchmen. The French were caught thus between the two and cut to pieces. This achieved, the victorious cavalry turned its attention to the French infantry, whom it took in flank.

Declining to make a stand, the enemy fled, and this, as far as the left and centre of the French Army, was the end of the battle. On the British left all was not yet over. There the cavalry, though they had

DETTINGEN

repulsed the French after so much strenuous fighting, had not yet done with their opponents, and pressed them again as hard as ever.

At this juncture the Royal Dragoons having disposed of the French infantry, who by this time had bolted, caught the remains of the French cavalry in flank, pressed thus as they were also in front. An utter rout followed and the entire French Army was speedily in headlong flight towards the two bridges at Seligenstadt, and also to seek any fords they could discover. Many plunged into the river haphazard and were drowned. There was no pursuit: there should no doubt have been. Stair proposed it and the King refused to permit it. Hence the remains of Grammont's force escaped unmolested. Why Noailles did not hurry up his men and take the British in the rear is a mystery; but as a matter of fact he did not, as far as can be ascertained, even cross the river during the battle. There he remained at Aschaffenburg without doing or attempting to do anything, and with him were some 20,000 men. The French losses amounted to some five thousand men, killed, wounded and prisoners. The loss of the British was 265 killed and 561 wounded and these casualties occurred mainly on the British left.

Several French standards and kettledrums were captured. Of the 7th Dragoons (The Queen's Own) Lieutenant Falconer, Cornet Hobey, one serjeant, ten privates, and twenty-two horses were killed. Lieutenant Frazer, Cornet St. Leger (who afterwards died of his wounds), one quartermaster, two serjeants, thirteen privates, and thirteen horses were wounded. General Clayton was also slain.

The Duke of Cumberland received a slight wound in the leg from a bullet. His horse, like that of King George, bolted, but in a different direction, carrying his rider into the middle of the French infantry. The Duke, however, managed to extricate himself from his somewhat dangerous and doubtless unpleasant predicament.

King George on the occasion displayed the greatest possible bravery and exposed himself recklessly. It is the fashion with some writers to laugh at his behaviour on the field at Dettingen, but of his genuine courage none can have any doubt. His remark when his runaway horse was captured and he had dismounted was characteristic. 'At least,' quoth he, 'I can trust my own legs not to carry me in the wrong direction.'

That night the Regiment remained on the field of battle, and unmolested by Noailles. On the morrow, leaving the wounded behind, it moved off to Hanau, and later encamped on the banks of the River Kinzig whither the Allied Army had already preceded it.

Early in August the King marched towards the Rhine, crossed it beyond Mainz, and advanced to Worms. The Regiment was then employed in West Germany, but there was no renewal of hostilities, Later in the autumn they repassed the Rhine and were marched back to Brabant and Flanders, where they went into the usual winter quarters. Meanwhile, the King had returned to England, soon to be followed by Lord Stair. This veteran officer had not been well treated, and had good cause to feel more than hurt. There is a very lengthy paper in which he expresses in the plainest terms and in no uncertain strain his disgust at the scanty courtesy meted out to him, and in which he announces his intention to resign. Truly he had a genuine grievance for was he not appointed to the supreme command, and despite that was he not prevented from conducting the campaign?

Still, such instances have occurred to other commanders before the days of Lord Stair, have occurred since, and probably will occur hereafter It is a curious reflection, but we were still not at war with France—for war was not formally declared until some months after the British, their Allies, and the French had been at hand-grips at Dettingen.

When Stair resigned—for nothing could shake his resolution so to do—General Wade, now for the purpose being promoted to the rank of Field-Marshal, was appointed to the Command. Failing Stair, he was perhaps the most fit or the least unfit person available for the post, but compared with Stair as a military commander he was a long way the inferior soldier.

It will here be fitting to quote the notice of this campaign for the year 1743, as it appears in the Manuscript Regimental Record.

1743. The Regiment embarked and arrived in Flanders and was present at the Battle of Dettingen 27th June new stile where it lost 17 Horses.

The next entry mentions that in 1744:

Capt. John Guerin of Sir John Cope's Dragoons serving in Flanders, his servants and baggage to be taken on board at Dover.

Extracts from the returns of the Killed and Wounded, printed in the *London Daily Post*, No. 2730:

> The Queen's Regiment of Dragoons, Lieut. Falconner, and Cornet Hobby, killed; Lieut. Fraser and Cornet St. Ledger, wounded. The loss of men and horses not separately stated, Cornet St. Ledger died of his wounds.

This is the whole of the information to be gathered from the Manuscript Regimental Record for the years 1743 and 1744, and is typical of the singularly slender contents of that volume.

Practically the one fact of importance gleaned is that at any rate Captain Guerin was not with the Regiment at Dettingen.

Unfortunately there are no means of ascertaining with any degree of accuracy what officers were really present at the battle.

It may also be remarked that it was in 1742 and not in 1743 that the Regiment embarked for Flanders. In the various accounts of the battle of Dettingen the valuable services of the British Artillery do not appear to be sufficiently recognised. It is, however, an undoubted fact that when on arrival—tardy, it is true, but unavoidably so—they got into action and silenced the guns of Noailles, or a portion of them, their excellent practice was of the utmost importance to the Allies. Let us consider the situation. Drawn up ever within range and for a protracted period, both the cavalry and infantry were exposed to a most heavy and galling fire, and a fire too that they were powerless to resist or to avoid, seeing that the river separated them from the French artillery. However heroically both the cavalry and infantry may have behaved, there is still room to award the just meed of praise to the British Artillery, and it should be un-grudgingly given.

From a letter to 'Mr. Urban,' the time-honoured nom de guerre of the Editor of the *Gentleman's Magazine*, vol. 14., p. 30, we are led in believe that considerable friction existed between the English and Hanoverians throughout the campaign, and that allegations of misconduct had been made against the latter at the battle. Those who are curious may consult that useful publication thereon. Mr. Urban's correspondent warmly defends the Hanoverian troops, and not by any means in the most polite terms.

In volume 13., pp. 434 and 435, there is a curious battle map, with a lengthy explanation, and stated to be 'Published by Authority,'

PLAN OF THE BATTLE OF DETTINGEN

THE BATTLE OF DETTINGEN.

References to the Allied *Army*.

A The Incampment of our Army from the 16th, 17th and 18th of *June*, as it came up succeffively to the 21th in the Morning, when it marched. The *Englifh* Troops being pofted near *Afchaffenbourg* at N° 1. the *Hanoverians* at N° 2. and the *Aufrians*, with their Right to the Woods and Marifhes, at N° 3.

B The March of the Army in two Columns, which halted at the Entrance of the Wood, where it drew up in Order of Battle, upon Advice that the *French* were pafsing the *Mayn* at *Selgenftadt*, their Infantry over two Bridges, and their Cavalry fording it.

C Difpofition of the Army before the Battle, including the Batteries E. which at firft made a continual Fire, and advanced towards the Enemy with the Lines about Noon, continuing to do fo till near three a Clock, when the Enemy retired. The Infantry is diftinguifhed by Colours (fmall) and the Cavalry by Standards (large).

D The Independant Companies in a Corner of the Wood, to cover our Baggage, which had retired behind it during the Battle.

F A Body of Cavalry obferving a Party of the *French* on the other Side of the *Mayn* at N° 10. who made a Shew as if they intended to ford the River.

G Three Batteries oppofed to thofe of the *French*, which they had erected on the other Side of the *Mayn* to gall our Troops in their March, and from the Firft of which, N° 11, at about eight a Clock in the Morning, they began to play upon our Rear, compofed of *Hanoverian* and fome *Englifh* Troops, having the King at their Head.

H The Village and Rivulet of *Dettingen*, which the *French* took Poffeffion of in order to attack us.

I The Attack of the *French* Houfhold Troops, which broke thro' the three firft Lines of our Infantry, but were repull'd with Lofs.

L The Attack of the *Fremo* Guards and other Infantry in Brigades, which took us in Flank, bur could not break thro' us, becaufe our Infantry of the Wing, altho' it be not marked here, formed immediately a Flank againft them, repulfed them, and forced thofe who did not throw themfelves into the River and fwim over, to retire along the River, to regain the Village of *Dettingen*.

M The March of our Army in Order of Battle purfuing the Enemy, in which the firft Line of the Cavalry was ord:red to take the Lead, and which kept before the Infantry till they arrived at the Place where the new Camp was marked out, and taken Poffeffion of.

N The new Camp where the Army paffed theNight after the Battle, and from whence it begin to march between nine and ten a Clock in the Morning, in order to reach the Camp between *Hanau* and *Franckfort*.

References to the French *Army*.

N° 4. The Camp of the *French* from the 16th and 17th of *June* to the 24th.

5. The Shifting of their Camp till the Day of Battle.

6. March of the *French* from Break of Day on the 16th of *June*.

7. March of five Brigades of *French* Infantry to attack the Head of the Bridge, and the Town too of *Afchaffenbourg*, which they took Poffeffion of finding no one there to oppofe them.

8. Paffage of their Infantry over two Bridges at *Selgenftadt*.

9. Paffage of their Cavalry at the Fords.

10. Other Fords, before which a Party of *French* Horfe were placed, N° 10.

11. The firft Battery of five Pieces of Cannon, from which the *French* play'd upon our Rear about eight in the Morning, which, (as has been before obferved) the King led on in Perfon

12. The fecond and third Batteries, which likewife took us in Flank during our March.

13. The fourth and fifth Batteries, which cannonaded our Army, whilft it was drawing up in Order of Battle, and which continued cannonading it during the Battle, and till the Action was near over

* Plain where the *French* drew up and advanced with the Battery of Cannon from which they only fired during the Engagement of the Houfhold Troops, and which difappeared immediately, upon their and other Troops, as well Horfe as Foot, being repulfed.

16. Retreat of the *French*.

† Their Incampment after they had reaffed the *Mayn* over the Bridges, and at the Fords, when a great Number were drowned, particularly of the *Freme* Guards. & groi *Veffitein*

N. B. *The Earl of Stair efcorted by thirty Dragoons going from the Camp at Afchaffenbourg to the other Side of the Main, in order to reconnoitre the Pofition of a Body of French that had advanced thither; a Party of them compofed of about fifty Men coming out of the Wood (4,) where they lay in Ambufh, attack'd the General's Efcort, and wounded in the Leg Mr Littleton his Aid de camp, the Earl of Stair himfelf having narrowly efcaped, a Ball which grazed his Head, and blew off his Hat.*

The map and explanation are here reproduced in facsimile.
From the *Gentleman's Magazine* we obtain a list of the French standards which were taken at the Battle of Dettingen. Details of six are given on p. 385, vol. 13.

1. A white Standard, finely embroidered with gold and silver, a thunderbolt in the middle, upon a blue and white ground. Motto, *Sensere Giantess*. Both sides the same.

2. A red Standard, two hands with a sword, with a laurel wreath, and imperial crown at top. Motto, *Incorrupta Fides et Avita Virtut*. On the other side the sun. Motto, *Nee pluribus impar*.

3. A yellow Standard, embroidered with gold and silver. The sun in the middle. No motto.

4. A green ditto in the same way.

5. The Mast of another tore off; but appears to have been red.

6. A white Standard, embroidered with gold and silver. In the middle a bunch of nine arrows tied with a wreath, all stained with blood; the Lance broke; Motto *Alterius Iovis altera Tela*.

This was the standard of the Black Musketeers. The cornet who carried it was buckled to his horse, and the standard to his body. A serjeant of Hawley's Dragoons captured it.

CHAPTER 7

Fontenoy
1744-1745

The victory at Dettingen had this effect; the French were unable to reinforce their army in Bohemia, and it was not long before they were ejected from Germany.

The scene of the future conflict was therefore limited to the Austrian Netherlands. Marshal Saxe was nominated Commander-in-Chief of the French Army in Flanders, and found himself the leader of a powerful and fully equipped force with which he opened the campaign quite early in the year, and by the beginning of April he had concentrated 80,000 men on the line between the Scheldt and the Sambre. To oppose him Marshal Wade, who, as has been told, had succeeded to the command on the resignation of Lord Stair, had some 55,000 men: a mixed army of British, Austrians, Dutch and Hanoverians, of which the proportion of British was about 21,000, or an increase of 5000 men over the numbers engaged in the preceding year. But it was no easy task to get this composite force, or rather its leaders, to act in hearty co-operation, In fact Saxe, the finer soldier by far of the two, not only possessed a much stronger force but had the advantage of a command which was not hampered by the jealousies and selfish intrigues which still further weakened the striking power of the so-called Allied Army.

The Dutch and Austrians were the chief trouble, for each desired and obstinately insisted on the protection as far as possible of their own frontiers, regardless of the general plan and conduct of the campaign as proposed by Wade on his arrival at the seat of war. But another serious drawback must be mentioned. Wade himself actually did not arrive until nearly six weeks after Saxe had concentrated his force. That Wade did not arrive was not, however, his fault.

It was not until the second week in May that he reached the Allied Army after a voyage retarded through contrary winds when he did actually embark, but a voyage at first delayed through the political events which were taking place in England; these, however, need not be particularised.

Reinforcements from England were sadly needed, yet they were not sent in any sufficient numbers. The Queen's Own Dragoons, however, appear to have been more lucky in this respect than some other corps. We read under date 5 March 1744/5:

3 officers, 40 men and 60 recruit horses of Cope's Dragoons to march from Daventry to Barnett (Barnet).

Also 7 March 1744/5:

3 officers, 38 men and 38 horses to march from Barnett to Gravesend and remain till they can embark for Ostend.

Again, on 25 March 1744/5:

20 men and 30 horses of Cope's Dragoons to march from Barnet to Gravesend and remain till they can embark for Ostend.

Cannon tells us that the Regiment was joined in the spring of 1745 by a number of men and horses, and this presumably refers to the drafts above named. He also tells us that the establishment of the Regiment had been augmented, but to what extent he does not state. Apparently the draft which actually went to Flanders amounted to three officers, fifty-eight men, and as many horses. The names of the officers again are not given.

The Allies concentrated in the neighbourhood of Brussels and had the pleasing prospect of endeavouring to hold not only Flanders and Brabant, but also Hainault and the line of the Sambre with a force far inferior to that of the enemy. For quite a month they remained in camp; meanwhile Saxe struck off to the west and then to the north, where he obtained possession of Ypres and Fort Knock, thus effectually blocking the Allies off from the harbour of Nieuport, which, with Ostend constituted their two bases of operations on the coast. The British now could only land troops and stores at Ostend. Here there were huge magazines of stores, guns, &c, and it was an open question whether Saxe could not obtain possession of that port as well. It was also quite possible for him to attack Bruges or Ghent or both, or he might besiege Tournai. Slowly the promised reinforcements reached Wade, and they had duly arrived by the first weeks of July. Meanwhile an important step had been taken. Prince Charles of Lorraine had crossed the Rhine, forcing his passage, and had occupied with his army a strategic position on the borders of Alsace. To him was now given the Command-in-Chief of the whole of the Allied forces.

A blow, and a serious one, could at this juncture have been struck at the French, and probably an energetic campaign could have been initiated, but it was not to be.

Prussia suddenly entered the field, and by threatening Bohemia with invasion caused the instant return of Prince Charles of Lorraine across the Rhine. Wade, who had already received from England most definite orders to immediately take the field and to act with energy, felt bound to obey, and after consultation with the other commanders, crossed the Scheldt, hoping thereby to come to hand-grips with Saxe, who was now comfortably entrenched on the river Lys between Menin and Courtrai. Having, however, crossed the Scheldt, the Allied Armies proceeded no further but lay for weeks in camp. Various other schemes were proposed which need not be here particularised, for one after the other either leave to attempt them was refused from England or the objections of one or other of the foreign commanders prevented their accomplishment.

Finally, a move was made to the south, and the Allies marched as far as the plains near Lille. Here, as before, a halt was called, and more was not attempted. Near Lille the army did not remain long, and it was soon quietly marching back again. Thus matters rested through the months of August and September, during which period Wade repeatedly but vainly proposed scheme after scheme for the prosecution of the campaign. By October the season was too advanced, and after this parody of campaigning the troops settled then into quarters for the winter.

Wade had had enough of it, and no wonder; so, resigning his command, he returned home. He was succeeded in the command by the Duke of Cumberland, who was also appointed Commander-in-Chief at home.

The new commander arrived in Flanders in April, succeeded in concentrating his troops at Soignies near Brussels by May 2, and actively took the field the next day, when his armies struck camp and marched off to the south.

Saxe meanwhile had as usual taken time by the forelock, and after a final march which appeared to threaten Mons, had turned off, and by 30 April had opened his trenches and was regularly besieging Tournai The movements of Saxe were swift and were executed with consummate skill. Through an extensive and masterly use of his powerful cavalry he so managed to conceal his real objective, Tournai, that no intelligence of his position or intentions was then received, neither did it become known to the Allies for some days. By 9 May Cumberland had arrived at Brussoel, having passed through Cambron, Maulbay, and Leuse. That evening he found himself face to face with Saxe.

Early on the morrow two troops of the Queen's Own Dragoons were employed in driving back the out-guards and piquets of the enemy, a prelude to the serious engagement which was to take place next day. Apparently there were no casualties—at least none are reported, nor indeed is the affair mentioned in the Manuscript Regimental Record, though it is duly recorded by Cannon in his History. These advanced posts of the French had been stationed in certain copses. A more particular account of the position here follows. The enemy were discovered drawn up in a very strong position. In their front the ground was much broken up by copses, small woods, and enclosed fields. These had been occupied by bodies of irregular troops, foreigners mainly. The employment of foreign irregular mercenary troops had been imitated by the French from the Austrians, who largely made use of Pandours and Grassins in their armies.

Pandours, it is stated, were so called because they were originally raised from the mountainous districts in Lower Hungary near a village called Pandur as Hungarian foot soldiers in the Austrian service. Campbell the poet mentions them thus:

When leagued oppression poured to northern wars
Her whiskered pandours and her fierce hussars.

The Grassins were a species of light militia, originally raised as the Arquebusiers de Grassin in 1744. Grassin was their first Lieut.-Colonel and had belonged to the Regiment of Picardy. Both Pan-

H. R. H. WILLIAM DUKE OF CUMBERLAND

MARSHAL DE SAXE

dours and Grassins were expert foragers, intelligence purveyors, and scouts. Their services were of the utmost use to the French armies as well as a constant source of irritation and annoyance to the British and their Allies. Behind these copses and fields the ground gently rose, and on the crown of hill was situated the village of Fontenoy. The centre of the French position was here, and the village was fortified.

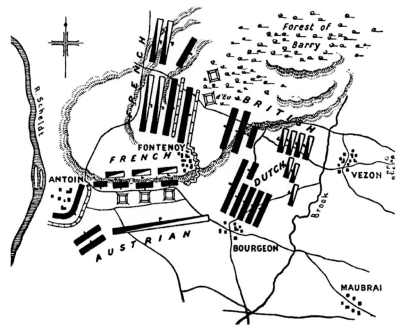

FONTENOY, 11TH MAY 1745

The right of Saxe's army rested on a village called Antoin, close to river Scheldt and due west of Fontenoy. Here there was a castle and some fortifications and also on the other side of the Scheldt a battery of guns. The left of Saxe's army extended almost at right angles to the line of the right as far as the Forest of Barry. Between Antoin and Fontenoy three strong redoubts had been erected, and between Fontenoy and the extreme left were two more, the nearer being known as the Redoubt d'Eu. Between Fontenoy and this redoubt there were also two lines of entrenchments. Roughly speaking, the French front occupied two and a half miles, and was, as will be understood, most strongly defended and as strongly held. It completely blocked the road to Tournai. In front were three villages, Vezon, Maubrai, and Bourgeon, which formed a triangle the apex of which, Bourgeon, was nearest to Fontenoy and distant about half a mile as the crow flies; the other two were about a mile apart, and were a mile and a quarter in the case of Vezon, and a mile and a half in that of Maubrai, from the same place. When the operations began, Vezon and the broken ground already mentioned were occupied, as has been said.

The night prior to the battle the Allies encamped at Maubrai. On the morrow the Pandours were dislodged, and as they retired to Fontenoy they set on fire the village of Bourgeon. Cumberland, accompanied by the Count of Königsegg, who was in command of the Austrians, and the Prince of Waldeck, who commanded the Dutch, then advanced to reconnoitre

The disposition of the French Army was as follows. The two redoubts on their left were armed with cannon and commanded the ground between the Forest of Barry and Fontenoy. The double line of entrenchments was held by nine and eleven battalions of infantry, behind which were two strong bodies of cavalry. Further on the left and behind the forest was more infantry and more cavalry. Fontenoy itself was strongly held and the three redoubts between that place and Antoin were armed with cannon and protected by infantry and a strong body of cavalry. The entire French force totalled about 56,000 men of all arms.

As a supporting force to the Pandours and Grassins in the village, the copses, and the fields, twelve squadrons of cavalry were drawn up in front of Fontenoy and part of the way up the slope. These retired when the French outposts were driven in.

It was decided to attack on the morrow, but not without a difference of opinion on the part of the Allied commanders. Cumberland and Waldeck advocated this course; Königsegg dissented, but was outvoted. The attack was determined on thus: the British right should attack the French left, the Dutch the centre, and the Austrians the French right. The balance of strength was in favour of the French, even without the adventitious advantage of redoubts and entrenchments, as the Allies could muster no more than 50,000 men. At 2 a.m. on the morning of 11 May the Allied Army set itself in motion, the British marching on Vezon, the advance being led by fifteen squadrons of cavalry under General Campbell. Three hours later Campbell's men had passed through Vezon and had formed on the ground beyond it. Their duty was to cover the British infantry while it was forming to attack the Redoubt d'Eu and the double line of entrenchments which lay between it and Fontenoy. To Brigadier Ingoldsby with the following troops was assigned the task of assaulting the Redoubt d'Eu: the 12th Foot, 13th Foot, 42nd Highlanders, a battalion of Hanoverians, and three six-pounders. These were detailed to carry the redoubt, and Cumberland had no doubt but that they would succeed while he and the main body of British infantry advanced and at the point of the bayonet, if need be, had the task of clearing the enemy out of the entrenchments. To reach these entrenchments a considerable stretch of open ground had be covered, and troops advancing were within range both of the Redoubt d'Eu and the guns at Fontenoy. On the appearance of Campbell's cavalry the French at once opened fire with their guns, and for an hoar the squadrons remained stationary and exposed to it. Campbell was then wounded—his leg was carried away by a cannon-shot. Nobody appears to have known what plan of action had been arranged for the cavalry, and the men, after being exposed uselessly, as it turned out, to a terrible gunfire, were later withdrawn.

All this time the British infantry were steadily forming beneath a hail of cannon shots under the actual command of Lord Ligonier. Six British guns now came up by the order of Cumberland and were posted on the extreme British right. Here they rapidly came into action, and it is recorded that one of the first shots killed the Duke of Grammont. The battle was at present an artillery duel, but the French paid far more attention to the ranks of British infantry than to silencing the British guns, and from d'Eu and Fontenoy a terrific fire was directed against them. Meanwhile, what was happening in the centre and on the left? The three attacks had been designed to be simultaneous. Ligonier was now ready and despatched a message to the Duke of Cumberland by his aide-de-camp, Captain Jeffrey Amherst. The Dutch and Austrians were also reported to be ready, and a general advance at once began.

We must now return to General Ingoldsby. His instructions, one would imagine were sufficiently clear to be incapable of misconstruction. He had but one objective, and that was the Redoubt d'Eu. This he was directed to assault and capture. What he actually did in that way was nothing, despite the fact that order after order was delivered to him from Cumberland, who was chafing at the delay. The result was that the redoubt, un-assailed, still continued to pour shot into the British ranks. At length Cumberland proceeded to make his frontal attack on the two lines of entrenchments, just as if the redoubt had been non-existent, and Ingoldsby a phantom. But how had the Dutch and Austrians fared meanwhile? Both advanced, but coming, as they speedily did, under a heavy fire not only from the village but also from the three redoubts on the French right and the battery across the Scheldt which flanked them, could not be persuaded to advance far. The gunfire prevailed; both Dutch and Austrians fell back and took what cover they could get, but, worse than all, a body of Dutch cavalry, panic-stricken, turned tail and bolted from the field, riding in their headlong flight right into some of the British cavalry, nor did they check their wild flight until Hal was reached.

This was the serious position on the left of the Allied Army. On the right the attack which Ingoldsby had been ordered to make on the Redoubt d'Eu, as has been already stated, never took place, despite the more than once repeated orders of Cumberland. Weary of waiting that General then initiated an attack on the entrenchments in his immediate front, without further delay. The British troops were drawn up in perfect lines, and their disposition from right to left was as follows: the first line: Grenadier Guards, Coldstream Guards, Scots Guards; the 1st, 21st, 31st, 8th, 25th, 33rd and 19th regiments; the second line: The Buffs, 23rd, 32nd, 11th, 28th, 34th and 20th regiments, with the addition of some Hanoverian battalions on the left. Each battalion was accompanied by two guns, to which the men harnessed themselves, and which by these means were dragged into position.

And now Ingoldsby, perceiving what was in progress, considered that the time had come to do something, though to obey orders even at the eleventh hour, and too late as it proved, does not appear to have occurred to him. What he did was this. He placed his troops in line with the advancing British, and thus the attack began.

Quietly and with as slow a step as if it were a parade movement the British lines moved on towards the entrenchments of whose existence they were fully aware but which they could not as yet see, as owing the formation of the ground and the fact that the entrenchments were over the crest of the slope they were invisible. For half a mile their path extended, flanked on the right by the Redoubt d'Eu and on the left by Fontenoy. From the guns of both these posts a constant and terrific fire was kept up which crashed continually through their ranks. Still they marched calmly on with drums beating and muskets shouldered, closing up their ranks as the murderous fire made gaps therein. Now it happened that the second line had a more extended front than the first, seeing which the Hanoverian troops on the extreme left were gradually relaxed in their speed and took up such a position as to enable them to march on in what practically became a third line. On towards the redoubt and the village the Allies steadily proceeded. And now the enemy's guns, which had been turned inwards first, played directly down the lines, doing fearful havoc, and later when the village had been passed actually took the advancing British and Hanoverians in the rear. More and more fell, but still at the same pace and for ever closing up and with a constantly decreasing front the Allies continued their steady advance. On arrival at the crest of the ridge the enemy at length came in view. They were posted behind a breastwork, distant some hundred yards. Onward the Allies still maintained their advance, though their front was now sadly contracted, so heavy had been their losses. At fifty yards distance from the breastwork it is stated that Lord Charles Hay of the Grenadier Guards bareheaded saluted the enemy and, producing his flask with somewhat, as it seems to us in these days, theatrical courtesy, drank to the foe, hoping, as he shouted, that the enemy would wait for them and not run as they had at Dettingen. He also added a few stirring words to his own regiment. The Guards answered with a cheer, which was echoed, though by no means lustily, from the ranks of the enemy, whose officers advanced to the front to acknowledge both toast and speech. By this time barely thirty yards separated the hostile lines, and then it was the enemy fired.

The time for the British had now arrived, and volley after volley was poured into the French. It is stated that no fewer than nineteen officers and 600 men of the French and Swiss Guards went down before the first fire. Another regiment, that of Courtin, was nearly exterminated, and the Aubeterre regiment lost nearly half its numbers. To sum up, the first French line was absolutely discomfited. And here it should be mentioned that the hand-dragged guns by frequent and well-timed discharges of grape-shot had played a most important part in this terrible struggle.

In vain Marshal Saxe had brought up additional troops to endeavour to stay the advance of the Allied forces. He had posted the regiments of Couronne and Soissons and the Brigade Royal behind the King's regiment and the French Guards, but without avail. The British volley-firing swept them away, and the Allies pressed on and on till they reached the French camp, which was situated about three hundred yards in the rear. It looked as if, despite their terrible losses, victory would declare itself on the side of the Allies. Ligonier certainly at the time esteemed the battle as won, and it is pretty nearly certain that his opponent Saxe agreed with him in that view.

At any rate, Saxe despatched a message to the King, who with his son the Dauphin was watching the course of the battle from a windmill near which their sumptuous tent was pitched, in which he suggested their retirement across the Scheldt without delay. Meanwhile Saxe busied himself in rallying his beaten or shaken troops.

The battle now entered upon a new phase. Cumberland, who evidently was ignorant of the failure of the Dutch attack on Fontenoy, when the first entrenchment was carried had detached two battalions from his left to, as he thought, assist in that attack by assailing Fontenoy and its defences on their right flank. Saxe, observing the movement and realising its danger, brought his cavalry into play, and by a charge of that arm these battalions were driven back.

One of these battalions then spread itself out along the left flank of the British. The effect of this movement, combined with the effect of the French guns, which had never ceased to play upon the Allies, was to press the British first into two columns and afterwards in into one, the two being joined so as to present the appearance of a large solid though elongated parallelogram. This body was attacked by the first line of the French cavalry, which had been fetched from the rear of the Forest of Barry and came on at a headlong gallop. The French charge was, as usual, met by volley-firing and was practically shattered. To these succeeded the second line, only to meet with similar failure.

A third charge, in which the Household Cavalry of France—

the celebrated Maison du Roi—took part, succeeded the first two failures, and like them was beaten back with terrible loss.

To return now to Cumberland and the Allied right, when we left in the French camp. Here the shot-torn ranks of the British were halted. Assistance, or rather support, from the Dutch under Waldeck was not forthcoming; for, as we know, his attack had failed. Lacking support, therefore, the British, who were still under the heavy cross-fire of artillery from the redoubts, slowly withdrew as far back as the crest of the ridge, where they were reformed with the intention of making a second attack. Cumberland, who had been in communication with Waldeck, on the promise of the latter to again endeavour to take Fontenoy, started for the second tine at the head of the British.

Marshal Saxe, however had not been idle. He had brought up his reserves which had been posted at Ramecroix and had also transferred the artillery from the French right to a position fronting the British advance. Among the reserves, it may be mentioned, were the celebrated Irish Brigade Saxe also had rallied the remnants of the French first line which had been driven earlier from the entrenchment.

Cumberland was therefore in the presence of three distinct artillery fires, and was faced not only by the rallied troops but by large reinforcements of fresh troops. In the same method again the British advanced and with almost the same success as at first, for they penetrated nearly as far into the enemy's camp as before. It seemed as if nothing, not even the Irish Brigade, could withstand the steady after volley which the brave, heavily stricken but devoted men poured into the enemy's ranks.

Cumberland was, however, completely outnumbered by this time, so heavy had been his losses; and a new danger now threatened him. On both flanks the French infantry closed in, and this, added to the unceasing cannonade on both flanks and in front, rendered a retreat a absolutely imperative. This retreat was not, however, in any sense a rout, but was conducted in the most orderly and regular manner, the Allies forming a square with their guns in the centre. As a preliminary Lord Ligonier despatched a couple of battalions to seize and hold the roads which led through to Vezon. Burning to achieve somewhat, for hitherto they had not gathered many laurels in the combat, the French Household Cavalry attempted to charge the rear just at the moment that the British faced about to retire. Their charge was met by the British Guards and some Hanoverians, who, turning again poured in effective volleys at close range on the advancing foe. The French suffered very heavily, and it is stated that every officer of one regiment was either killed or wounded, while another regiment was well-nigh exterminated. As will be seen, Fontenoy was as much an infantry battle as Dettingen was a cavalry combat; in fact, the British cavalry after clearing the coppices as has been already narrated, had during the engagement nothing to do. With the retirement, however, matters changed, and such few squadrons as were within call, among them being the Horse Guards Blue, hurried to cover the retreat, passing en route through the terrific cross-fire of the enemy's guns. The retreat continued at as slow a pace as the advance, broken only by the facing about of battalion after battalion at short intervals. Thus did the shattered remnant of the British to the end keep their pursuers at bay till Aeth was reached, beneath the guns of which place the wearied troops found shelter. Details as to the action of the British cavalry do not appear to be forthcoming. We are told that the total loss of this arm of the service in the battle 'exceeded three hundred men and six hundred horses.' The Horse Guards Blue and the Royal Dragoons suffered most.

The casualties in the ranks of the 7th (Queen's Own) were ten men and forty-six horses killed; Lieut.-Colonel Erskine, Captain-Lieutenant Ogilvy, Lieutenant Forbes, Cornet Maitland, Quartermaster Smith, thirty-five men, and forty-seven horses wounded; one man and two horses missing. This is the account given by Cannon in his History.

The Manuscript Regimental Record, however, differs, and we will give here the entire entry concerning the campaign, battle and casualties:

1745
New Stile
11th May. The Regiment was present at the Battle of Fontenoy where it had 67 horses killed and the following officers killed and wounded

Cornet Potts	Killed
Lieut.-Colonel Erskine	Wounded
Capt.-Lt. Ogilvie	Wounded
Cornet Maitland	Wounded
Quartermaster Forbes	Wounded
Quartermaster Smith	Wounded

An extract from the Daily Post of Monday, 13 May 1745, is appended to this account, which confirms the death of Cornet Potts and also states the numbers of men and horses killed, wounded, and missing to be the same as given in Cannon.

It may be remarked that Lieutenant Forbes was also Quartermaster. Clearly the number of horses (sixty-seven) stated to have been killed must have been an error on the part of the man who wrote up the Regimental Record.

As will be seen from the above narrative of Fontenoy, the details concerning the Regiment are few indeed. Search has been made in many directions to obtain facts to supplement this account, but without avail. It would seem that nobody concerned cared to keep any detailed record. Mr. Skrine, in his interesting and careful work, Fontenoy of the Austrian Succession, has apparently found the same difficulty and is unable to assist us in any way. For the purposes of this history it was, however, needful to give an account of this celebrated battle, and why that account should prove to be so abridged has now been explained. It must also be remembered that a regimental history is in no sense a history of the British Army as a whole.

The official account given in the Gentleman's Magazine (vol. 15., p. 246) furnishes one or two other particulars. We learn that by the order of His Royal Highness, General Ligonier 'caused seven pieces of cannon to advance at the head of the Brigade of Guards, which soon silenced the moving batteries of the enemy.' The hostile guns in the entrenchments and redoubts are not of course referred to here. When the order to retire was given, the two regiments selected to hold the road were Lieut.-General Howard's and the Highlanders. The first was posted in the churchyard of Vezon and the second 'in the hedges where they had been, posted the day before.' The baggage of His Royal Highness was sent off to Aeth at about 2 p.m. During the action it had remained at Brussoel. Marshal Königsegg, who had been hurt by a fall from his horse, also went to Aeth, but not until the retreating allies had passed through the defiles. He arrived there in the evening, but His Royal Highness remained with the troops and did not reach Aeth till 3 a.m. on the morrow. No colours, standards or drums were lost, and one standard was taken from the enemy. A gun was abandoned in consequence of the early flight of the contractors with the artillery horses.

CHAPTER 8

Campaigns After Fontenoy
1745-1749

As has been already stated, the Allies made good an orderly retreat to Aeth. Here they rested for a space of five days, a period which was employed in reorganising the regiments which had suffered so heavily in the glorious but disastrous battle of Fontenoy. That there was no panic is witnessed by the calm tone of the General Orders. Returns of the killed and wounded were called for and obtained, the sale of the effects of deceased soldiers was ordered, and widows—for as of yore the wives followed the armies into the field—were sent back home.

The loss in horses too was duly registered. It appears that some of the officers had engaged French deserters as servants, and these, as possible spies, were ordered to be dismissed and furnished with passes to Brussels.

One man, by name Patrick Crowe, who was suspected of espionage and was stated to have come from Bavaria, was seized, summarily tried, and hanged. An officer of one of the cavalry regiments was accused of cowardice and tried by court-martial. For the honour of the service we are glad to record that he was acquitted, being able to prove in the most positive manner that his flight was involuntary, he being carried away by the cavalry panic which affected the Dutch cavalry and from which he was totally unable to extricate himself.

Meanwhile the enemy were not inactive, and it soon became apparent that Aeth was by no means an ideal rallying-point for Cumberland's army. Round and round the camp the Grassins and light troops of the enemy incessantly hovered, now cutting off stragglers and at times convoys. Four days after the battle it became needful for Cumberland to take active measures to rid the outskirts of the camp from this annoyance, and a force of 300 horse and 600 foot was detached under Brigadier Skelton to effect this purpose. Henceforth the Pandours and Grassins therefore were kept at a distance.

On 16 May Cumberland put his troops in motion and advanced to Lessines, which he reached after crossing the river Denaire, near that place, by means of a pontoon bridge. From this place some of the British regiments which had suffered most severely were sent into garrison, their places being taken by others that had not been so terribly cut up. At Lessines considerable reinforcements of foot joined the army from various quarters. Barrell's, now the 4th Foot, was the first to arrive, Fleming's (the 36th) arrived from Oudenarde, Ponsonby's (the 37th) from Bruges, Beauclerk's (the 31st) and Ligonier's (the 48th) from Ostend. Dutch regiments were also at once despatched by the States General.

There was, however, a great dearth, as usual, of British troops at home, and in consequence six battalions of Austrian infantry and twelve troops of Hussars were taken into British pay by agreement with the Empress Maria Theresa in the month of June.

The Austrian contingent amounted actually to 5600 men— 8000 had been promised but were not supplied. From the Landgrave of Hesse also 6000 men were similarly obtained. By 11 June Cumberland could muster some 16,000 infantry and 4200 cavalry in his right wing, and this force he speedily organised into a condition of efficiency. The total of the Allied Army was now, professedly at least, no less than 70,000 men, but in reality it did not nearly amount to this number.

The capture of Tournai was now the immediate object of Saxe. He remained with his army safely entrenched on the field of Fontenoy the night after the battle. On the morrow he despatched all the available Grassins to pursue and harass the rear of the retreating Allies. By these a dozen guns were captured and a considerable quantity of stores for which there was no transport.

One detachment penetrated as far as Leuze, gathering in by the way some three thousand stragglers. The wounded to the number of 1200 which had been left at the Château de Brussoel were captured, stripped and robbed by the Grassins, who, not content with this, served the surgeons in charge in a similar manner. This act of barbarity led to an interchange of letters between Cumberland and Saxe, in which the recriminations were mutual.

Meanwhile Tournai still held out despite the strenuous efforts of Löwendahl, the French commander, who was entrusted with the task of capturing it. During the day on which Fontenoy was fought he pressed his attack, especially directing it on the horn-work which defended the Gate of Seven Fountains. On 15 May this defence was carried and on the 18th a practicable breach was made in the bastion close by. The garrison determined to capitulate. Negotiations were entered into with Löwendahl. It would appear that the Governor, by name General Baron Van Dort, and all principal officers under his command, with the single exception of Lieut.-General Leuwe, were determined on yielding the stronghold. The matter was referred to the States General, who returned two different decisions to Marshal Königsegg, their aged commander-in-chief. One paper authorised a surrender on the best terms obtainable, the other urged that the place should be defended as long as possible. As will be seen, the decision rested, or rather was laid, on the shoulders of the gouty and aged veteran.

He called a meeting of the other commanders to his bedside, where it was unanimously determined to hold out.

Instructions were therefore sent to adopt the second course. The admission of wounded men or any other useless encumbrances into the citadel of Tournai was forbidden, and an attempt to cut through by the besieged cavalry in the direction of Oudenarde was suggested, failing which they could at need fall back on their horses for food. Frequent sorties to annoy and retard the enemy as long as possible were similarly enjoined. It was, however, of no avail; the garrison were resolved to surrender, and the capitulation was signed on 20 June 1745. It was as a matter of fact a disgraceful affair. The surrendered garrison were conducted to Ghent.

The loss of Tournai had most disastrous results on the relations existing in the wings of the Allied Army. Bickering, nay even actual quarrels, broke out between British and Dutch soldiers, so much so that none of the latter were permitted to enter Ghent citadel with side-arms, and all were rigidly excluded therefrom after tattoo.

The release of the besieging force gave a considerable accession of strength to the army of Saxe, and it was not long before he took advantage of it to resume the offensive.

On 30 June he despatched a force of three battalions and thirty-four squadrons under the Marquis de Clermont-Gallerande to occupy Binche. This place lay half-way between Mons and Charleroi, and consequently threatened both places. Now the possession of Mons and Charleroi was very dear to our Dutch Allies, and despite all that Cumberland could urge they took measures to secure them. To effect this the left wing of the army was weakened to a dangerous degree. Eight battalions were hurriedly sent to Mons, six to Namur, and detachments also thrown into Charleroi and Aeth. Cumberland had now hardly more than 40,000 men with which to take the field. The Dutch next urged a withdrawal from Lessines, declaring that it was impossible to hold it. Cumberland, though very unwilling, was practically forced to acquiesce, and on the day the French occupied Binche the Allied Army was withdrawn to Grammont. How great an error this was may be understood by what follows.

At Lessines the French would have been compelled to remain behind the Scheldt or else to have fought on ground chosen by the Allies, and on ground which was most favourable for cavalry action. Naturally this retreat delighted the French.

In vain Cumberland, who still possessed 40,000 British and Hanoverian troops, urged an attack. His actual proposition was to intermingle the troops of the four allied nations and then attack the enemy, a plan which of course differed considerably from that which had been employed at Fontenoy. The Dutch, however, would have none of it. And now the French advanced, after demolishing the works of Tournai—works which had been designed by the celebrated Vauban. In six columns they marched on Leuze, pushing on the Grassins as far as Lessines, where they threatened the rear of the retreating army.

On the morrow the French advanced to within a league of Grammont and seemed about to attack. Cumberland then threw six British battalions with twelve guns into the place. Next day, after manoeuvring for ten hours, the enemy encamped about Grammont at about a league's distance, but on the other side of the river.

It now remained for Cumberland to decide on one of two

courses. He must abandon either Flanders or the city of Brussels. Still he had some hopes of being able to save Ghent, and for this purpose despatched Lieut.-General Moltke to seize Alost with ten squadrons of cavalry and three battalions of infantry, the instructions to that commander being to watch the enemy and if necessary to throw himself into Ghent.

Moltke occupied Alost on 8 July. On the same day Saxe sent 15,000 men under Du Chayla in the direction of Ghent, with orders to reconnoitre the base of the Allies.

The Grassins moved eastwards, and, seizing the Château deMassenew about eight miles from Alost, entrenched themselves there.

Moltke heard of the move and despatched the Royals (1st Foot) to dislodge them. The Royals surrounded the place. Fifteen volunteers were called for by Grassin, the colonel who was in command, to carry news to Du Chayla. Of these, five succeeded in getting through Du Chayla then threw his remaining troops across the embankment leading to Ghent and awaited the coming of Moltke. The Royals were unable to take the Château, being unprovided with guns, and rejoined.

Moltke proceeded to Ghent as instructed. On arrival near Melle Priory he came in touch with the enemy, of which a brigade was drawn up across the embankment, and the hedges and houses on either side were filled with troops. There was also a battery of twelve guns about forty yards on the left. The Royals, who led the column, charged the enemy in their front and routed them. They then captured the French guns, but it happened that the pontoons of the enemy had been posted in rear of the guns and behind them were two French brigades. During the infantry combat the British Dragoons (Rich's) and the Hanoverian cavalry made a dash for Ghent. They, however, found themselves opposed by a small party of the Berry cavalry under a Captain St. Sauveur, who in a most gallant manner succeeded in holding them for a sufficient time to allow of the arrival of another French brigade.

Moltke's force was now in a very tight place. It was outnumbered about three to one. Still the little force pressed on towards Ghent, despite the heavy fire poured into them on either side. With a loss of one-third of their number they, the Royals, Rich's Dragoons, and the Hanoverian cavalry at length reached their destination. Brigadier Bligh, who commanded his own regiment, Handyside's, and some Dutch cavalry managed to fall back on Alost, gaining that place trough the woods and lanes; but with the loss of his tents and baggage, which were captured by the ubiquitous Grassins. From Alost he retreated to Termonde. His loss was 292 of the British. Fifteen hundred prisoners were taken by Du Chayla and a great deal of booty. At least 500 of the Allies fell, added to which Cumberland's despatches were captured. The enemy, however, did not escape without considerable loss, as nearly 1000 were either killed or wounded.

This disaster sealed the fate of Ghent. Saxe reinforced Du Chayla by 15,000 men under Löwendahl and the place was at once invested.

On 11 July it was summoned to surrender. The French Dragoons succeeded in swimming the moat unopposed and at the same time the Bruges Gate was seized by two battalions of grenadiers. The garrison, Dutch and British, the latter being the Welsh Fusiliers and the remains off Moltke's force, retired into the citadel. Moltke himself escaped at the head of the Hanoverian Cavalry. At Sluys he arrived that night, but was refused admission into the place. He wished to pass through and on to Antwerp. As the Dutch would not permit him to enter the gates, the unfortunate men, weary and war-worn, had to make their way to Ostend.

The citadel of Ghent was now besieged, the investment beginning on 14 July. The defenders do not appear to have offered any resistance of a strenuous nature. The batteries were about to open when the Dutch governor gave in, and, practically without striking a blow in defence of the place, surrendered.

A huge accumulation of military stores; the Dutch garrison, the Royal Welsh Fusiliers, and a number of Rich's Dragoons found themselves prisoners of war. The one redeeming exploit of this disaster was the gallant conduct of the quartermaster of Rich's Dragoons, by name Kelly. He managed to slip out of a sally-port before the capitulation was signed, at the head of four cornets and 160 men, and safely reached Antwerp. On arrival at the camp of the Allies it is pleasant to read that he was rewarded with a cornetcy.

Cumberland's position was now the reverse of pleasant. The enemy advanced on Mons and it became impossible for the Allies to continue at Grammont. They accordingly on 10 June retreated to Meerbeck near Ninove. The news of the fall of Ghent now arrived.

Cumberland at once retreated to Brussels, and thence a few days later to Dieghem. On 24 July the enemy under the personal command of the King (Louis XV) advanced to Oordeghem, thus threatening to cut Cumberland off from Antwerp. In consequence he retired two days later to Saventhem just outside Brussels on the south-east.

Cumberland's army was now in a position of considerable danger, and there is no doubt that he fully appreciated the difficulties of his situation. To keep the field he had no more than 34,000 men, while the enemy possessed a force of 70,000 troops concentrated at the King's camp, to which must be added a body of about 8000 men who were under the command of the Comte de Clermont-Gallerande. If Brussels fell it would be the immediate precursor of a terrible disaster.

If Antwerp was captured his communications with Holland would be cut and all means of subsistence for his army would be gone, which would mean starvation and surrender. Added to which the condition of Ostend was far from satisfactory. At this juncture Cumberland urged most strongly that of the two evils the loss of Antwerp would be the greater. His representations alarmed the British Government, who despatched, in the first place, Major-General Braddock to report on the condition of Ostend, to examine its defences, and at the same time they forwarded to that place an immense quantity of stores. Cumberland's proposal had been to cut the dykes, but this course was vetoed by the Austrian Minister in Flanders, by name Wenzelius, who later became Prince von Kaunitz. It remained therefore for the British to defend the place as best they might.

To add to the troubles of the British Government, at this juncture the rebellion in Scotland known as 'the 45' now broke out. Prince Charles Edward landed near Moidart on 19 July 1745; but definite news of this event did not reach London for nearly a month.

This rendered the preservation of Ostend a matter of vital importance, as free communication with a seaport became an absolute necessity. A fleet was despatched to cruise off Ostend. The garrison there was reinforced by two regiments, viz. the Royal Scots Fusiliers and the 32nd Foot. An Austrian General by name Chanclos was sent to examine the defences of Ostend. His verdict was that the place might be held, but that a larger garrison was needed.

Accordingly a combined battalion of the Guards was sent thither from London. On 7 August a fleet of British and Dutch transports entered the harbour with a large quantity of artillery and stores of all kinds. Cumberland two days later despatched thither from Antwerp the 18th Royal Irish. These defensive measures were taken none too soon. On the day of the arrival of the Royal Irish Löwendahl appeared on the land side with a force of 21,000 men and 6000 horses laden with fascines. He began operations by summoning Fort Plassendael to surrender, and the summons was obeyed. Owing to the refusal of the Austrians to permit the dykes to be cut the enemy were able to erect batteries on the shore, and these rendered the approach of the British warships well-nigh impossible.

Meanwhile Cumberland had removed his headquarters from Saventhem to Vilvorden, a place on the Antwerp-Brussels Canal and situated about seven miles north of the last-named city. Here he encamped in a line along the canal. His position was therefore somewhat nearer Antwerp, and he at the same time did not leave Brussels absolutely uncovered.

Towards the north as far as the canal-head on the Rupel the British and Hanoverians were posted, while the Dutch occupied the remainder in the direction of Brussels. Along this line a chain of redoubts and entrenchments was erected with all speed, and here Cumberland remained on the alert. In the army itself he had many disciplinary troubles. Desertion was rife, and various other military crimes such as marauding and robbery were of only too frequent occurrence ; it is needless to add that, as was then the custom, offenders when detected were punished with the utmost severity. And now the main French Army under the immediate eyes of both Louis XV and Marshal Saxe put itself in motion. An advance was made to Alost, by which both Brussels and Antwerp were threatened. Their right was extended as far as Termonde, at that time held by a Dutch garrison and the British 48th Foot. Cumberland in vain endeavoured to increase the garrison by sending a reinforcement of some 600 men by boat from Antwerp, but the attempt failed, and unfortunately many lives were lost.

Saxe then laid regular siege to Termonde. Around the town the ground had been flooded, but the Marshal succeeded in draining off the water. After a twenty-four hours' bombardment the place surrendered on 13 August, and the captured guns were forthwith despatched to increase the artillery already employed in the projected attack on Ostend.

The siege of that place began regularly on the following day. A lucky and, it may be said, very plucky attempt on the part of the navy succeeded in removing such British cavalry as happened to be shut up in the fortress and, what was more to the purpose, managed to land as well a considerable addition to the stores besides certain reinforcements. The completion, however, of the heavy French battery on the sandhills near the shore prevented entirely any repetition of the exploit.

The bombardment now began, and was carried on continuously. General Chanclos, who was in command, had a task which was well-nigh impossible of performance in directing the defence. His Dutch troops were by no means in a state of discipline. True he had five British battalions to stiffen his force, but it will hardly be believed, though true, that these British were there shut up in the besieged fortress without one general officer of their own nationality. An attempt by Cumberland to send thither Lord Crawford and Brigadier Mordaunt at the eleventh hour was prevented by the French shore-battery aforementioned.

Meanwhile the main French Army drew nearer and nearer to Cumberland's position on the Antwerp-Brussels Canal. Louis took up his quarters at Lippeloo, Saxe at Opdorp, and their right wing extended as far as Steinhuffel. The course of events in Scotland now caused the British Government to suggest the despatch of troops from the seat of war in Flanders to that country for the purpose of home defence. To do this at the time was impossible. An agreement was, however, entered into with the Dutch Government to despatch to England the 6000 men which they were by treaty compelled to do.

George II was himself out of England at the time, though he was endeavouring to return there. Contrary winds, unfortunately, detained him at Helvoetsluys.

Ostend was now very hardly pressed. On 22 August, after a night attack, the besiegers succeeded in occupying the covered way, though not without suffering great loss. Two days later the place surrendered. By the articles of capitulation the garrison was permitted to evacuate Ostend with all the honours of war, and was then to be escorted to Austrian territory. Aeth, the sole remaining fortress in Austrian hands, was the next place to be besieged. After an eight days' bombardment it fell on 9 October. The garrison marched out with full military honours and joined the army of Cumberland. British troops were now returning to England in numbers. Cumberland soon followed them. Marshal Königsegg set forth for Vienna and Prince Waldeck took over the command of the remains of the Allied Army. These he as speedily as possible located in winter quarters.

Orders now arrived to send to England eighteen squadrons of British cavalry, four companies of artillery, and the field-train which had been quartered in Antwerp.

Such are the main incidents of this most disastrous campaign.

We must now state in order the few facts connected with this campaign which concern the 7th (Queen's Own) Dragoons. After the capture of Tournai the Queen's Own accompanied Cumberland's army and was encamped far some time near Brussels. The movements of the army in question have already been narrated.

After the fall of Ostend the Regiment went into winter quarters and there remained until February 1746. In that month orders were received for them to return to England. Accordingly they marched to Williamstadt and embarked on the transports which had been provided for them. As usual, contrary winds prevailed and the troops were delayed—one transport it is stated was stranded. The continuance of the bad weather then led to a change of plan. The men and horses were all disembarked and ordered to await a favourable opportunity for their voyage. Meanwhile the back of the rebellion had been broken, and the Government came to the conclusion that the presence of additional British troops in England was no longer needful, while their presence in Flanders was of greater importance. Accordingly the order for re-embarkation was countermanded. But though the Queen's Own Dragoons had not since Fontenoy been engaged in any actual active warfare, the Regiment would appear to have been considerably under its war strength. Accordingly, in March 1746 we find drafts ordered to be sent from home to the Regiment in Flanders.

On 25 March we read as follows:

102 men with 156 horses belonging to Lieut.-General Sir John Cope's Dragoons to march from Daventry to Gravesend, there to remain till they can embark on board the transports for Flanders.

This is succeeded by another order dated 28 March:

The men, recruits, and recruit horses of Cope's Dragoons upon their arrival at Barnet to halt till further orders.

On the next day we read that:

Three officers and 55 men with 120 recruit horses of Cope's Dragoons to march from Barnet to Dartford and remain till they can embark at Gravesend for Flanders.

The date on which the embarkation took place and the date of the arrival of this draft in Flanders we are not told. As will be understood from the facts narrated above, the opportunities offered to the cavalry to gain distinction in the campaign of 1745 were few; but the following year, though uniformly disastrous to the arms of the Allies as a whole, was from a cavalry standpoint one in which that branch of the service at any rate acquired distinction.

During the winter Saxe, until January 1746, had his quarters at Ghent. How his time was employed in that city is well known and need not be entered upon in detail. Waldeck lay at Malines with his army. Contrary to usual custom, however, Saxe towards the end of the month initiated a winter campaign. Brussels was invested on 30 January and capitulated after a siege of three weeks, Waldeck being quite unable with the forces at his disposal to either retard the catastrophe or to relieve the place. In consequence no fewer than 15,000 men—the garrison—became prisoners of war. Next Vilvorden fell, and with it practically the most important stores and artillery train of the Dutch became the spoils of the victors. Louis was now disposed to rest on the laurels his armies had gained in Flanders, and even opened negotiations for peace with Maria Theresa. These negotiations for peace, however, came to naught, the Empress declining to discontinue the war. Culloden had been fought and won; the rebellion in Scotland had been stamped out, and the British King was at length able to devote his energies to Flanders.

Following on the refusal of Maria Theresa to negotiate with Louis came the despatch of 50,000 Austrians under the command of Prince Charles of Lorraine across the Rhine. Saxe was again despatched to the seat of war as commander-in-chief—he had been in Paris for some time, where as a victorious commander he had been duly fêted.

His first act was to lay siege to Antwerp, and it surrendered on 31 May. Waldeck retreated to Breda from Malines and there entrenched himself while awaiting the advent of Lorraine. From England now came considerable reinforcements: troops which had been released owing to the crushing of the rebels in Scotland.

Six thousand Hessians and thirteen thousand Hanoverians reached the camp at Breda during the month of May. Thither also at the end of June came four British cavalry regiments, among them being the 7th Queen's Own, and six regiments of infantry the force being under the command of Sir John Ligonier. Waldeck's army now numbered about 50,000 men, and, more than this, the Austrians under Lorraine had so far advanced as to be within four days' march. As to the events of this year, however, the manuscripts are silent. Ligonier landed at Williamstadt as the only available spot, since Ostend and Antwerp were both in the hands of the enemy. His infantry was not, however, in a condition to take the field, being destitute of powder, artillery horses, and baggage wagons. Added to this, the relations between the British and Austrian commanders were, to say the least of it, strained. Saxe entered upon another period of activity. He detached the Prince of Conti to reduce Mons and St. Ghislain, and both these places fell. That commander then proceeded to lay siege to Charleroi.

Evidently Maestricht and Namur were the objective of the French campaign. It was not, however, until July that the Allies were able to move. In that month they shifted their camp to Terheyden, and here they remained until 17 July. At length an advance was made with a view to relieve Charleroi. Saxe, who had been lying with his main army near Antwerp, at once marched to the Dyle between Aerschot and Louvain. Lorraine after a terrible march arrived at Peer, then crossing the Demer at Hasselt in a southerly direction reached Borchloen on 27 July. It appeared now as if Namur could be saved and possibly Charleroi as well.

On that very day, however. Charleroi surrendered. A complicated series of manoeuvres now took place. At times the opposing armies were almost within musket-shot of one another, at far apart. Battle was more than once offered and refused—nay, more, the opportunities for attack which were given on either side were disregarded. Muy fell to Saxe. Lack of forage and subsistence in of the Allies was severely felt. Namur was besieged by the and capitulated, after a siege of eleven days, to the Prince of Clermont. The Allies then fell back in order to protect Liège, crossing the river Jaar on 7 October. They took up a position admittedly bad: the river Meuse was in their rear and Saxe with an overwhelming force in their front and steadily approaching them with an evident intention of fighting. He crossed the Jaar on 10 October, and that night the Allies and the enemy faced one another with the prospect of a battle on the morrow.

The position was as follows: The Allies faced due west and were drawn up across two roads, both of which led to Liège. Their right rested on the river Jaar, where it was covered by three villages, Sluys, Fexhe, and Enick, which were occupied by the Austrians, who had strongly entrenched themselves. On the open plain to the south of Enick and stretching as far as a village named Liers the four British battalions and the Hanoverians were posted. On their left and in rear of Liers were the Hessians; next came the Hanoverian cavalry, who extended southwards as far as the village of Voroux; then came the 6th Dragoons and the 7th Queen's Own Dragoons, who filled in the interval between Voroux and Roucoux; lastly, the Dutch troops occupied the ground between Roucoux and Ance. As a position it was decidedly bad. Not only was it possible and indeed comparatively easy for the enemy to turn the left flank, but if the right flank were turned the line of retreat to Maestricht would be cut. Added to these, the Meuse ran in the rear and the presence of two ravines between the rivers Jaar and Melaigne in the centre allowed but one narrow means of communication between the right and left of the Allies.

Lorraine as well as Ligonier both appreciated the difficulties of the case. The former, however, held his right with as much strength as he could to avert the danger on that flank; and so matters rested on the night of 10 October. On the morrow came news that the French were masters of Liege. This compelled the Prince of Waldeck to denude his left of eight battalions from Roucoux and to post them on his left flank with cavalry supports to guard against any attack in that direction. Hence to the British, Hanoverians, and Hessians alone was left the defence of the three villages, Roucoux, Liers and Voroux, and they numbered but eight battalions. Obviously the left was the weakest, and on that flank Saxe at once advanced. He was superior both in numbers and in strategical position. His army amounted to 100,000 men, while that of the Allies was not exceeding 80,000, and of these fully 26,000 were safe behind the Austrian entrenchments on the extreme Allied right near the river Jaar.

In dense columns he first assailed the Dutch on the Allied left, following this up by an almost simultaneous attack on the centre in which fifty-eight battalions in three columns were launched upon the three villages. Splendidly did British, Hanoverians, and Hessians answer to the call. More than one attack was repulsed, but each time Saxe sent forward fresh troops, and these at length prevailed, so much so that Roucoux and Voroux were perforce abandoned, though the enemy could not prevail at Liers. Ligonier, man of action as he was, rallied his men, and brought them back to the attack with the result that both Roucoux and Voroux were recaptured. Meanwhile the Dutch had lost heavily and were retiring from the left across the rear of the position. Then it was that Ligonier brought the British cavalry into play. First using them as a check to the enemy in order to permit the Dutch to make good their retreat, he next ordered the withdrawal of the British infantry, employing his cavalry in a similar manner as before. Having possessed himself of the villages, Saxe then ordered a general advance, practically a pursuit of the retiring British. On through the captured villages the enemy's battalions made their way, but they were not for long unchecked. At them were launched the Royal Scots Greys, the Inniskillings, and the 7th Queen's Own Dragoons. The advancing enemy wavered at the onset, their ranks were broken, and before the charge of the British cavalry they turned and fled. This gallant charge without a doubt was the main cause of the retreat being accomplished without a great disaster.

Ligonier in a letter thus briefly describes the action of the cavalry:

. . . . the enemy's foot with great shouts began to come out upon the plain in great numbers, but on our marching upon 'em with the Cavalry at a great trot, sword in hand, they run back into the villages much faster than they come on.

There are no records of the loss of the 7th Queen's Own (if any) on this occasion. The History of the Royal Scots Greys does not mention Roucoux. The casualties among the Inniskillings amounted to three officers wounded, six men killed and seven wounded, one man missing, and fourteen horses killed, wounded, or missing. Cannon in his History of the 7th (Queen's Own) gives a brief account of the battle of Roucoux; the Manuscript Regimental Record, however, does not mention either the battle or the campaign. In the upshot the British, Hanoverians, Hessians and Dutch crossed the Meuse by means of three pontoon bridges at Visé. The Austrians, who had been inactive during the combat. partly from the configuration of the ground and partly from the position of the French containing force, also retired in good order.

The Allied forces halted at Maestricht, and it being now the season for winter quarters they so disposed themselves in the country along the lower Meuse, Breda being occupied by the British.

The campaign of 1747 began slowly. In January a convention was signed with Holland by which that country agreed to furnish 40,000 men and to pay two-thirds of their cost. Hanover promised 16,400, Austria 60,000, and the British contingent numbered 13,800. On paper this amounted to 130,000 men, but as a matter of fact not more than 112,000 were forthcoming when in February the Allies concentrated at Breda.

The plan of the French was to carry war into Holland, and for this purpose two armies were formed: one, under Löwendahl, was destined to wage war in Dutch Flanders and to lay siege to Sluys and Port Phillipine; the other, commanded by Saxe, who was now styled Marshal-General, would threaten Southern Holland and lay siege to Maestricht. Löwendahl succeeded in his designs: west Flanders being secured rendered the left of Saxe safe from any hostile movement.

The Dutch were in a state of terror, and a revolution took place which ended in the election as Hereditary Stadtholder of Prince William of Orange-Nassau, the son-in-law of George II. Meanwhile the Allies marched on Antwerp with the intention of besieging it.

Löwendahl frustrated their design by throwing a powerful force into its citadel. In the event the Allies abandoned their intention and retired to a position in which they covered Maestricht. Here they suffered much loss of strength, as desertion was rife and, what was even worse, dysentery had broken out. Still Cumberland had even yet an army of 81,000 men with 250 guns, and ever and forcibly urged the active prosecution of the campaign. But here he found himself opposed by the commanders of the Dutch and Austrian contingents, and, outvoted at council, he had perforce to remain idle.

Louis XV was now expected to join the headquarters of his army, That army had been resting quietly on the borders of Hainault awaiting him. He reached Brussels on 31 May and reviewed his host—a host which numbered 140,000 men. On 22 June Saxe put his army in motion and advanced to Maestricht. Cumberland simultaneously endeavoured to shift his quarters to Tongres. On arrival there he found Saxe already in possession of the position and had in consequence to fall back on Maestricht. For the space of a week the opposing armies were occupied in manoeuvring to obtain a strategic advantage one of the other.

The last design of the Allies was to seize Herdeeren Heights, and thither they proceeded, only to find that again the wily Marshal-General had forestalled them; he had left ten battalions at Tongres, and with 120,000 men had mainly advanced and seized the Heights, thus placing himself between Cumberland and Maestricht. On 2 July the opposing armies joined battle. This battle is for some reason or another blessed with no fewer than six names. In various authors it figures as Val, Vol, Keselt, Latal and Laffeldt.

The Allies were drawn up in three lines, the Austrians on the right, the Dutch in the centre and the British Hanoverians and Hessians on the left. Laffeldt was situated in their front. Saxe drew up his infantry in two lines on the higher ground with his cavalry in the plain beneath.

Louis in person, as at Fontenoy, was an interested spectator of the battle. Laffeldt, the key to the position, was first attacked. It was held by the British and German infantry. Three times the attack was successful and three times was the place recaptured, though fresh regiments sent up by Saxe were employed on each occasion. The Marshal-General made a fourth and this time a personal endeavour. At the head of the Regiment Le Roi and accompanied by six infantry brigades he returned to the charge; nor was this all, for he concentrated on Laffeldt the fire of many guns, the Royal Vaisseau and the Irish Brigade acting as supports to the movement. To assist the battalions in Laffeldt, who could with difficulty sustain this heavy attack, or rather who were unable to eject the enemy from the place in which they had managed at length to secure a footing, Cumberland brought up the whole of his left wing. Four of the infantry brigades of the enemy, however, took them in flank and they were driven back in some disorder.

Things looked badly at this juncture, but undismayed Cumberland ordered the Dutch cavalry to charge. Saxe perceiving them in motion, ordered a counter cavalry attack on the advancing Dutch. The latter did not await the onset but fled incontinently, and moreover in their flight threw into disorder five battalions of infantry that had been ordered up to recapture Laffeldt. On rode the French cavalry and unopposed pierced the centre of the Allies. This decided, or all but decided, the fortune of the day. A retreat began. The left and centre, hard pressed by the enemy's cavalry, moved off the

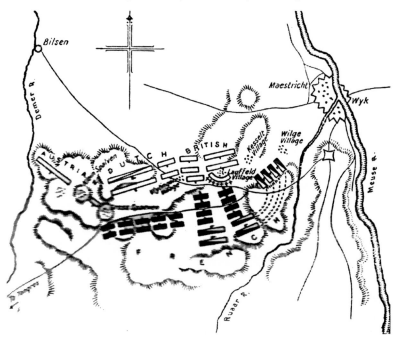

VAL OF LAUFFELD

field, but few indeed of them would have reached the River Meuse had not the veteran Ligonier, taking in the situation at a glance, given orders for a charge of British cavalry and thereby saving the Allies from total destruction.

Placing himself at the head of the Royal North British Dragoons, Rich's Dragoons, Rothe's Dragoons and the 7th Queen's Own Dragoons, he attacked the pursuing French cavalry. His charge was not to be withstood. The hitherto victorious French were rolled back in confusion and five of their standards were captured. Ligonier himself was taken prisoner in the mêlée, his horse having been killed. The charge saved the Allies. They succeeded in passing across the Meuse by pontoon bridges as after Roucoux, and reaching Maestricht encamped beneath its walls.

The Austrians as if it were Roucoux over again were not engaged seriously. They made good their retreat in a northerly direction. In this fight the Allies lost 6000 men and 16 guns, while 2000 men were taken prisoners.

Cannon tells us with regard to this year that the Regiment was encamped for a short time in the spring near the banks of the Scheldt, and was subsequently employed in operations on the Great Nethe and the Demer, during which a draft of fifty men and remounts of one hundred and twenty horses joined from England. The Manuscript Regimental Record for the year merely states that:

The Regiment was present at the Battle of Latal, Sunday 2nd
July (New Stile).

Cannon tells us, and the Record corroborates him, that the Regiment 'lost several men and horses on this occasion, and had Cornet Bulmere, five men and ten horses captured by the enemy.' The actual number of wounded were eight men and four horses, and four horses were killed.

After the battle, Saxe returned to his previous quarters at Tongres. His next move was to detach a French force to capture Bergen op Zoom, in which he was again successful, Löwendahl being as usual in charge of the siege.

Negotiations for peace were now again mooted; and indeed as far back as 1746 endeavours in that direction had been begun, but only to be broken off. The meetings began in January 1748, and, as usual in such cases, the conflicting interests of the various States concerned were hardly conducive to unanimity of action. April arrived, and with it Saxe again took the field. This time his cherished design of capturing Maestricht was to be attempted to be put into execution. It was so attempted and it was successful. The siege began on 15 April, Maestricht capitulated on 10 May.

This concluded the long-drawn-out struggle. The Treaty of Aix-la-Chapelle was signed by Great Britain, France and Holland on 18 October, 1748.

CHAPTER 9

Home Service
1748-1760

When hostilities ceased for the winter of 1747 the Regiment was stationed in the province of Limburg, and remained there facing the enemy, but on the defensive practically until the preliminaries of the Treaty of Aix-la-Chapelle were agreed upon during the spring of 1748. This agreement caused a suspension of hostilities, and the strain of the situation being thus relaxed, the British troops were scattered in cantonments among the peasantry inhabiting various Dutch villages. On the signature of the Treaty of Aix-la-Chapelle, which took place on 18 October 1748, orders were received for the immediate return of the British troops to their native shores.

Accordingly the 7th Queen's Dragoons marched to the coast. The exact port of embarkation does not apparently seem to be known, nor the exact date thereof; neither are any details of the voyage forthcoming. From the Manuscript Regimental Record, however, we learn that the 7th Queen's Dragoons on landing at Purfleet on 26 January were ordered to march to Avely, North and South Ockendon, and certain places adjacent, and after a halt of one or two nights to proceed as follows: One troop each to Colchester, Witham and Kelvedon, Chelmsford, Ingatestone, Brentwood, and Romford. A detachment which had been left behind in Flanders under the command of Lieutenant Lumsden arrived at Lowestoft on 3 March, and received orders to march to Yarmouth. Three days later this detachment was ordered to march thence to Colchester. The Regiment was now reduced to a peace establishment. This reduction took place between the dates of 6 February and 14 February.

On 14 February their quarters were shifted, four troops proceeding to Norwich and two to Yarmouth. A troop of the Regiment, however, mysteriously turns up on 21 February at Colchester from Chelmsford, having been despatched :hither to aid the civil power in the preservation of peace. Cannon now tells us that detachments of the Regiment were sent (places unnamed) on coast duty—that is to say, against smugglers. On 14 March the Troop or Detachment at Colchester was ordered to proceed to Yarmouth. The Regimental Record is here certainly rather vague.

Cannon is silent as to the events of the remainder of the year. The Regimental Record, however, tells us that 'the out-party of the Queen's Royal Regiment of Dragoons on the smuggling duty' was ordered on 14 November 1749 to join their troops, and that the Regiment was then to march from their present quarters to Exeter. The exact intended stations of the Regiment originally are not given, but from an entry dated 21 December we read that 'the two troops at Sherbourne (on the march to Exeter) to detach an escort with a deserter, on his way to London.' The next entry does not assist us. It is dated 1749-50, but no month is given or day specified, though it must have been before April:

> The 6 Troops of the Queen's Royal Dragoons, now at
> Dorchester, notwithstanding former orders to march
> Viz. 3 Troops to Wells, 2 to Frome and 1 to Shepton Mallet.

On 10 April 1750 the Regiment was ordered to concentrate at Wells, there to be reviewed by Lieut.-General Campbell, and afterwards to return to their former quarters.

The country was at this time in a very disturbed state. In several districts serious rioting had taken place. Around Trowbridge, Bradford and Melksham 'many hundred weavers having assembled in a riotous and tumultuous manner,' two troops of the Regiment were sent thither on 18 April to assist in preserving the peace. At Birmingham, later in the year, two troops of Honywood's Dragoons were most hastily sent to quell what was rumoured to be 'a great insurrection.' It was, of course, the outcome of the distress caused by the war which produced these disorders. It is seldom that during the war itself civil troubles are manifested in this country, but with a peace, and especially if a not remarkably honourable peace, the shoe begins to pinch; the excitement of the war is over, and sufferers have only their grievances and their poverty to occupy their minds. Hence then came the generality of civil troubles requiring armed intervention in those police-less days.

We have before spoken of the hardships inflicted on disbanded or reduced regiments. In the Gentleman's Magazine for 1749, page 132, a poem entitled A Reduced Officer's Complaint appears. It is too long to quote, but those curious may well refer to it.

It may here be stated that the distinguished regiment of dragoons known as The Duke's, and originally The Duke of Kingston's, despite all its gallantry and services in Flanders, was just at this period ruthlessly disbanded. Side by side with the Queen's Royal Dragoons they had more than once fought, especially in the last campaign.

To return to the subject of the disaffected districts. Some of the proceedings of the riotous weavers savoured of the comic. Having constructed several effigies of notable personages—one of these in fact representing the King—they proceeded to shoot them, and when considered sufficiently wounded or dead the dummy figures were decapitated. In various parts of England culprits were tried, but for a wonder the juries were lenient, few were convicted, in some cases juries disagreed and in others justice was professedly satisfied by binding the misguided men over to keep the peace.

How long the Regiment remained in quarters, or the date when the two troops rejoined, we cannot learn. This we do know, that some time between 18 April and 29 September the horses were at grass, for under the latter date we read 'the 6 troops of Cope's Dragoons (a return here to the old style) to march fourteen days after the horses are taken from grass to Gloucester, there to be reviewed by Lieut.-General Campbell.' This review took place presumably on 17 or 18 October, as after the review the Regiment received orders to march as follows : three troops to Worcester, one to Pershore, and two to remain at Gloucester. For six months we again lose sight of the Regiment, till on 13 April 1751 they received orders to concentrate at Gloucester again to be reviewed. The review took place on 26 April, the reviewing officer being Lieut.-General Onslow. After the review they, as usual, returned to their former quarters.

On 1 July 1751 King George II issued the dress warrant of that year relative to colours, clothing, &c, of the army. As this will be fully discussed in the special chapter on Uniforms and Equipment, there is no need to include it here, save in one respect. By this warrant it was directed that the 'number of the rank or seniority of the regiment' was to be borne on the colours and appointments from this date. The Regiment, however, did not at once conform to the order as to the number 7.

On 18 September we find them still styling themselves Cope's Dragoons On this date their horses were at grass, and they were ordered to march fourteen days after the animals had been taken from grass to the following places: Two troops to Shrewsbury and one each to Stafford, Burton-on-Trent, Walsall and Ashburn. These orders were countermanded on 10 October, and the new quarters of the Regiment were fixed as four troops at Worcester and one each at Pershore and Bromsgrove. Here they were to remain until such time as was needed for the whole Regiment to concentrate at Birmingham by 2 November. At Birmingham they were to be reviewed by Sir Philip Honywood, after which they were ordered to take the route given in the orders of 18 September. At the last moment, however, the troop for Ashburn was ordered to remain at Birmingham.

The colliers at Pembroke, Haverfordwest, and the adjacent places were now giving trouble, rioting in order to prevent the importation of corn. In consequence of this, the two troops from Shrewsbury were despatched thither on 29 February; the Birmingham troop was on the same day sent to replace one of the troops withdrawn from Shrewsbury.

On 9 April the troops were under orders to march 'so as to arrive at Lichfield on Monday the 21st.' At Lichfield they were to be reviewed by Lieut.-General Campbell, after which they were to return to their former quarters. On 14 April two troops detached at Pembroke and Haverfordwest were ordered to march immediately to Lichfield. The entire Regiment left Lichfield on 25 April and went into quarters as before. On 12 October the Queen's Dragoons were ordered to assemble at Stafford by 28 October, there to remain until further orders. Here they remained until ordered to Manchester on 10 April 1753, to be reviewed by Major-General Cholmondeley. Until 4 October the Regiment remained in Man-

1751

chester, Bradford, and the adjacent places. They were then ordered to proceed to Berwick-on-Tweed, where further orders as to their destination in Scotland were to be awaited.

On 25 October, when they had reached Durham, an order came directing them to halt until Major-General Cholmondeley should arrive to review them, after which they were to proceed according to their former orders. This is the last entry for 1753, but a note in faded red ink in the Manuscript Record states that:

> The numbers of the Regiments were first used in official orders and books in this year.'

We have no information as to the exact date upon which the Regiment entered Scotland, as the next entry vaguely states that in April 1754 all the six troops were at Perth. On 1 December we find one troop at each of the following places: Coldstream, Kelso, Dunbar, Dalkeith, Musselburgh, and Haddington; so far the Regimental Record, and Cannon's History gives us no more.

Dated from the War Office, 1 September 1756, we get the following letter:

> *Sir,*—I am to signify to you, It is His Majesty's Pleasure, that you make strict enquiry into the conduct of such of the private men belonging to the Regiment of Dragoons under your command, as have been in the foot service, and that you return to me the names of three of those men per troop, or the number of twenty-one from the Regiment., (if so many are to be found,) whom you can recommend as proper persons to be made serjeants in the new levies.
> I am, Sir, your most humble servant,
> *Barrington*
> To the Officer Commanding
> Lt.-Gen. Sir John Cope's Regt. of Dragoons
> At Haddington

From another War Office letter, dated 25 January 1757, we gather that several private dragoons from the Regiment were accordingly turned over to Major-General Holmes's and Colonel Leighton's regiments of foot (2nd Battalions) to be made serjeants and that £5 per man was ordered to be paid by the foot regiments in question to Sir John Cope's Dragoons. The information derived from these two documents, which were unexpectedly discovered, is certainly of interest.

Great Britain was now engaged again in a war with France, the immediate cause of which was a quarrel with that nation on the subject of the boundaries between the possessions of Great Britain and France in America. In consequence of this war the establishment of the 7th Queen's Own Dragoons was increased by 347 officers and men. In the struggle in America the Regiment was not called upon to engage. But on 29 January 1756 an important addition was made to the strength of the Regiment in the form of a Light Troop. This was raised by royal warrant, and consisted of one captain, one lieutenant, one cornet, one quarter-master, two sergeants, three corporals, two drummers and 60 light dragoons. Cannon states that the number of Dragoons was 63. The name of the Captain was William Erskine, who had previously held the commission of Captain-Lieutenant. His date of commission was 25 December. It is noted that both officers and men were mounted on small horses. Cannon tells us, but gives no date, that the Light Troop 'was subsequently augmented to upwards of one hundred officers and men.'

It will be remembered that in the year 1756 British Dragoons were first equipped as Light Dragoons. The first regiment to be so equipped was Elliot's, now the 15th Hussars; Burgoyne's followed—a regiment which that unfortunate and much-maligned officer was given the permission to raise 'on his own particular plan,' and of which the establishment presents some variations from that of other regiments

It is not generally known that for no less than seven years Burgoyne, who had left the British Army with a captain's commission and was residing in enforced exile on the continent owing to lack of means, was engaged in the careful and serious study of the methods then in vogue in the Light Cavalry of European nations, and that his subsequent successes at Belle Isle and in the Portugal-Spanish campaign few years later with his newly-raised regiment (now the 16th Lancers) ere the outcome of his thorough acquaintance with the subject. We shall later narrate in full, as far as the details are obtainable, the adventures and services of the 7th (Light Troop) of the 7th (Queen's Own) Dragoons. To return to the Regiment as a whole. In August 1756 we find that four troops were stationed at Musselburgh, two at Dunbar, and one at Haddington. In May the following year the Regiment was concentrated at Perth.

Between 1 September and 6 September we know that four troops were at Musselburgh, and one each at Dundee, Alloa, and Cupar. On 6 September three troops were moved to Dalkeith and one each to Kelso, Haddington, Dunbar, and Linlithgow.

By 3 March 1758 the various troops had been shifted as follows: two troops at Dunbar, two at Haddington and Coldstream, and three at Musselburgh, Leith, and Dalkeith, as on that date they received orders to march thence to Berwick and Kelso en route for England, their destinations being four troops to York, one to North Malton, and two to Pontefract and Ferry Bridge. This march was timed to begin on the 14th, 15th, and 17th of April, and they were to reach the above-named places on the 26th and 29th of the same month and 3 May respectively.

On 25 April the Light Troop received orders that on arrival at York it should forthwith proceed to Winchester so as to arrive at that city on Wednesday 24 May. On the same date the remaining six troops were ordered south and stationed in Essex as follows: three troops at Chelmsford and one each at Maldon, Brentwood, and Romford. Their march was to be timed so as to reach these places on 7 June. The Light Troop, however, did not proceed to Winchester, as we read that these orders were cancelled two days later. It was sent instead to Petersfield, there to encamp on 13 May. On 19 May the Maldon troop was sent to Ingatestone. On 20 May the Light Troop received orders to hold themselves in readiness to embark at Portsmouth on foreign service.

The destination of this troop was France, where it formed part of an expeditionary force despatched to that country. On 7 June five troops of the Regiment were ordered to assemble at Chelmsford and one troop, that at Ingatestone, to remain at its quarters. A review was to be held at Chelmsford by Colonel Douglas, after which the various troops were to return whence they came.

On 3 July all the six troops were despatched to stations nearer London, viz. two troops to Epping, two to Hoddesdon and Hatfield, one to Barnet and Ricksend (Rickmansworth?), and one to Hampstead and Highgate. On 26 July the six troops were ordered to

London to be reviewed and subsequently to march, four troops to Canterbury, one to Sandwich, and one to Faversham and Ospernage (probably Ospringe near Faversham). On 28 September two troops were ordered to Barnet, two to St. Albans, and two to Enfield. The Light Troop had now returned to England, and, having disembarked at Portsmouth, was on the same date ordered to Hackney.

On 14 October the entire Regiment was removed to new quarters, three troops proceeding to Chelmsford, one to Witham and Braintree, one to Ingatestone and Chipping Ongar, one to Brentwood, and the Light Troop to Romford. One of the Chelmsford troops was on 7 November shifted to Braintree and Bocking.

And here, if we may be pardoned a slight digression, we will narrate two really curious and interesting military episodes which had taken place in the locality during the first half of the seventeenth century. Our reason for this digression is the fact that the circumstances to which we shall allude are practically unknown to the general military public, but are surely of considerable interest to military readers as throwing side-lights on the conditions obtaining between soldier and civilian in those far remote times. Our authority in both cases is to be found in the State Papers, Domestic Series, 1628 to 1640.

In 1628 Buckingham had brought over from Ireland a strong force of Irish infantry, which he proceeded to billet by companies in various towns and villages in England. These troops were generally alluded to by the polite designation of Buckingham's Blackguards. Now the circumstances, though they led to most deplorable results, were such that there was much to be said on both sides. These men were practically unpaid, as their slender pay was most often months and months in arrear. The billeting money too was, if paid, insufficient to satisfy the requirements either of the men or the townsfolk upon whom they were billeted; but as, like the pay, it was ever far in arrear, the results may easily be conjectured. Having had the company commanded by a certain Captain Roy (or Roys) Carew (or Carey) billeted upon them, the inhabitants of Maldon exercised a privilege then common to everybody and frequently exercised. This was to forward a petition to the Privy Council for the removal of their unwelcome visitors. Thousands of petitions of this kind are extant, frequently on the most trivial subjects. The petition is headed in form 'The Poor Distressed Inhabitants of Maldon.' They complain therein of the 'insolences and outrages' of the troops and state that the ' burden is intolerable.' They protest that the men 'command in our houses as if they were our lords and we their slaves, enforcing us and ours to attend them at their pleasure and to do the basest offices for them.' They state that 'not content with the diet proportional to the King's pay, they compel what they will and pay with violence.' The inhabitants declare that they are forced to stay at home to protect their houses and to guard their wives, daughters, and maidservants from outrage. The Council was about to grant their petition and to remove Captain Carey's company to Witham, a neighbouring place, which the worthy burgesses of Maldon had slyly suggested was far better able to sustain troops than they were. The people of Witham heard of this and promptly lodged a counter-petition, in which, among other things, they suggested that the men of Maldon had on previous occasions ill-used soldiers billeted there. The Deputy-Lieutenant, through whom the petition was forwarded and who probably wrote it, added this cogent reason for the non-relief of Maldon: 'If (he wrote) 'Maldon be freed from soldiers the Duke of Buckingham will find them (the soldiers) beaten where soever they are billeted.' The Maldon petition was accordingly refused.

On St. Patrick's Day 1628 the storm burst. This we learn from a despatch sent to the Council by the Deputy-Lieutenant on the morrow. He states that he had that day removed the men from Maldon to Witham on his own initiative, and for this reason: There had been a serious riot, in fact a small battle, at Maldon on the occasion of the Saint's day, in which no fewer than thirty men were either killed or wounded and Captain Roys Carew himself shot in the head; he adds, 'it is thought dangerously.'

The ostensible immediate cause of the riot was the tying of 'red crosses to the whipping-post and to a dog's tail.' We are not told whether they were the cross of St. George or that of St. Patrick. The townsfolk took up their arms, the country folk flocked in from the villages near with what weapons they could muster, and in the sleepy long street of Maldon a pitched battle raged for some hours. On the arrival of the soldiers from Maldon at Witham, similar scenes were witnessed, for the men of Witham rose in a tumult. The Deputy-Lieutenant endeavoured to pacify the enraged inhabitants and to get them to disperse. This he at last accomplished, but not until two soldiers were shot absolutely in his presence and he had been compelled to disarm the whole company, this being insisted on by the inhabitants. Now these riots were, one would suppose, so serious as to merit at least an inquiry. Nothing, however, was done. Nobody was arrested, nor as a matter of fact was anything further heard of the matter. The other case, which was an outcome of what was known as the Altar-rails Question, took place in 1640. Throughout the Eastern Counties there was a strong leaven of puritanism with which even the soldiers were infected. A company of foot under the command of a Captain Rolleston was billeted at Braintree. The parson of Bocking—an adjoining village connected with Braintree by one long street—whose usual style is, as is that of the parson of Battle in Sussex, Dean' presented these men with fifty shillings and a barrel of beer, possibly to ingratiate himself with them, he being a Laudian, pro-altar-rail cleric. The beer was consumed and the fifty shillings as well in all probability. More merry than wise, the men proceeded forthwith to Bocking, where, instead of cheering the Dean, they broke into the church, tore up the altar-rails, bore them off to Braintree in triumph, and made a bonfire of them before their captain's lodgings. Next they proceeded to Radwinter Church, where they tore down a statue of Christ and some heads of cherubim and seraphim, which they carried to Maldon and burnt in the high street, 'all as profane as the Sons of Belial.' On this occasion, however, the ringleaders were arrested and lodged in Chelmsford Gaol. Captain Rolleston in his report mentions a curious fact: that his men were 'most frequent in desiring their officers to take the Holy Communion with them.' Such things happened in the good old times': how different matters are now!

To return to the movements of the Regiment. On 6 June 1759 the various troops were shifted from their former quarters to Ipswich (three), Colchester (two), Sudbury (one), and the Light Troop from Romford to Yarmouth. 7 July, the Light Troop was ordered from Yarmouth to Maldon. On 31 July we find four troops at Chelmsford and Maldon, one at Ingatestone and Chipping Ongar, one at Romford, and one at Epping and Waltham Abbey. A month later, one of the Chelmsford troops was despatched to Bocking and Braintree.

On 2 November the first division of three troops of the Regiment ordered to assemble on 6 November at Newington, Islington, Tottenham, and Edmonton, and thence to march, one troop to Dunstable and two to Aylesbury. The 2nd division of four troops was to assemble at Chelmsford on 5 November and to proceed as follows: one troop to Amersham Wendover, and Risborough; one to Tring and Ivinghoe, and one to Berkhamstead and Hampstead: the Light Troop being sent to Hitchin and Baldock.

There is now a gap of over a year in the Regimental Record; the next entry being dated 24 November 1759, on which day 'one of the troops at Aylesbury' was ordered to march to High Wycombe and Beaconsfield. The Adjutant must have been a very idle man as regards keeping his books posted, for we find no other entry until 5 February 1760, when a detachment of six sergeants and 54 private dragoons were ordered to Wendover so as to arrive on 17 February. We are not told, however, where they marched from. The next entry is puzzling, as we have no clue to its origin:

11 Feb.—The drafts from the 7th Dragoons at Wendover to march to Watford.

British troops were now about to be sent again to Germany on active service, and on 10 March we find that six troops were ordered to march so as to arrive on 26 March at Greenwich and Blackheath, Wandsworth, Fulham and Putney, and Kingston, preparatory to their embarkation. This order was followed on the next day by a change of route. One troop to Watford, two to Aylesbury, two to Dunstable, and one to Edgeware and Stanmore. This entry is headed 'until they begin the march ordered yesterday.' Two orders are dated 26 March:

The two troops ordered to Wandsworth, Fulham and Putney to proceed to Blackheath, Gravesend, Lewisham and Lee preparatory for their embarkation to Germany.
The two troops ordered to Kingston to proceed to Wandsworth, Putney, Fulham and Clapham preparatory to their embarkation to Germany.

What became of the other two troops is not mentioned, nor is the station of the Light Troop given-Cannon says that the six heavy troops of the 7th Dragoons went to Germany, but the Regimental Record only records the departure of four. Possibly, however, this discrepancy can be explained later.

We will now narrate the story of the two expeditions to the coast of France in which the Light Troop of the 7th Queen's Royal Dragoons took part. These expeditions were the idea of Pitt, who thought that by making a descent in force on the coast of France he would attain one or both of two objects:

1. To draw off from either America or Germany any additional forces which were likely to be sent thither by the enemy.

2. To do as much damage as possible to the coast towns.

A costly armament was therefore equipped, consisting of thirteen battalions of infantry, three companies of artillery, a siege-train, and a composite force of cavalry under Elliot, which was made up of the Light Troops of nine dragoon regiments, the Light Troop of the 7th Queen's Royal Dragoons being one of the troops thus employed. The entire force numbered some 13,000 men. Charles third Duke of Marlborough was placed in supreme command, with Lord George Sackville as his second. A powerful fleet acted as escort to the transports, the latter numbering no fewer than 100 vessels. This large fleet set sail from St. Helens on 1 June, ad arrived at Cancalle Bay, an indentation in the coast about eight miles east of St. Malo, on Monday, 5 June, where by the evening of the day the whole force had landed. The disembarkation was but slightly opposed by the guns of a French battery which had been erected to protect the bay. The casualties only amounted to three seamen. Obviously it was easy for the guns of so large a fleet to silence the battery, and this they promptly did. Leaving a brigade of infantry to guard the landing-place the remainder of the troops marched forthwith on St. Malo. Here an attack was found impracticable unless by means of a regular siege, as the fortress and town were too strong. The Duke made certain dispositions to begin an investment in due form, but on receiving intelligence that the enemy were hastily and in great strength marching against him on every side, and that his retreat to Cancalle Bay would inevitably be cut off, promptly retired again to his ships. St. Malo, however, did not escape scot free. Under cover of night the Light Dragoons succeeded in reaching the harbour, where they burnt many of the naval stores, a man-of-war of fifty guns, one of thirty-six, all the privateers lying in the port, some of which had thirty, and several twenty and eighteen guns, in all numbering upwards of 100 vessels. It may be remarked too that during the execution of this damage the Light Dragoons were throughout under the very guns of St. Malo.

On arrival at Cancalle Bay Marlborough found Commodore Howe fully prepared to re-embark the troops, so much so that four brigades and ten companies of Grenadiers were embarked in less than seven hours and unmolested by the enemy. By 12 June the entire force was again on board the ships.

A private letter from an officer on board the Speedwell, one of the men-of-war, a sixteen-gun brig, tells us that one transport was lost between Jersey and Sark through running on a submerged rock, but that there was no loss of life. The inhabitants of Cancalle appear to have fled, and to have left the troops in quiet possession of the town. One regiment was sent a day's march into the interior to the town of Dol, about fourteen miles from St. Malo, where they were well received. No opposition was met with on the march thither, and it was there reported that there were not 500 French troops in that part of the country at all. St. Malo, he states, was found to be very strong and its walls both lofty and extremely thick, so much so that at least a month would have been occupied in reducing the place.

There were two strong batteries on the sea-front; the entrance to the harbour was both very narrow and dangerous, and could not well be entered by the British war-ships without very considerable risk of loss. Owing to contrary winds the fleet and transports were delayed in Cancalle Bay.

From another letter we learn that when the Regiment was sent to Dol, a part of the Light Dragoons accompanied them and advanced beyond that place. Here they fell in with the vedettes of a French camp. These fell back at once. The Light Dragoons gave chase, and after a long pursuit succeeded in taking two prisoners, whom they brought into camp. When the army was mustered on board the transports some thirty men were found to be missing, having been left behind. Several of these were afterwards brought off in French fishing-boats and exchanged for French prisoners to an equal number. The fate of the residue was never ascertained.

On the 14th there was some talk of disembarking the troops

again, or at any rate a portion of them, and the Grenadiers and Guards were told to hold themselves in readiness for such duty and to complete their ammunition. The town of Granville was then reconnoitred on 14 June, but was found unworthy of attention. Two days later the fleet sailed, and after beating about against the wind till evening was compelled to anchor off St. Malo. The wind rose, some of the fleet drove and others parted their anchors. Next morning all the ships returned to Cancalle Bay.

Until the 22nd the weather continued tempestuous, but at length the vessels were enabled to sail. On the morrow they passed Jersey and Guernsey, on the 25th the Isle of Wight was sighted, on the 26th the Bind veered round and carried them back to the French coast near Havre de Grace. Here a landing was at first designed, and the flat-bottomed boats were hoisted out for that purpose, but towards evening the wind became so boisterous that the boats were quickly hoisted in again, and the fleet stood out to sea. Next morning the wind moderated and the vessels again ran in to within a few leagues of the shore. The Duke of Marlborough then went in a cutter to reconnoitre, having given orders for four days' provisions to be served out to the troops in anticipation of a landing. Nothing, however, was done either on that day or the next.

On the 29th the fleet bore away to Cherbourg and anchored about two miles from the town. Some of the transports that lay nearer in were fired on from five or six different batteries, but no harm was done. Bodies of armed men were seen drawn up on the shore, some of whom appeared to be regular troops. In the evening orders were given for landing, but a contrary wind arising, it was impossible to execute them. On the morrow the fleet and transports weighed anchor and stood for England. arriving in the evening of the next day at St. Helens: so ended the first of the costly raids on the French coast. The second raid against the French coast left England on 1 August. This time it was under the command of Lieut.-General Bligh, the fleet being commanded by Commodore Howe.

The fleet arrived before Cherbourg on the 6th and proceeded to bombard the town. Early on the 7th they set sail for the bay of St. Marais, situated six miles from Cherbourg.

Here a landing was effected under cover of the fire from the frigates and bomb ketches, and despite the opposition of a large body of the enemy who were prepared to receive them. On the 8th, Cherbourg surrendered at discretion, as the enemy marched out and abandoned the place on the approach of the British. General Bligh then took possession of the forts of Querqueville, Hornet, and la Galette. Next day the preparations began for destroying the basin then being constructed and two piers at the entrance to the harbour. There were about twenty-seven of the enemy's ships in the harbour and thirty pieces of brass cannon were captured. Cherbourg, as far as its basin, piers, harbour, batteries, forts, magazines, and stores were concerned, was then destroyed, and smaller places near were similarly treated. To all this destruction the enemy were either unable or unwilling to offer any opposition.

The British force then re-embarked, carrying with them twenty-two fine brass cannon and two brass mortars; 173 iron cannon and three iron mortars were destroyed. Again tempestuous weather caused a long delay. It was not until 3 September that the fleet anchored in the bay of St. Lunaire, situated about twelve miles east of St. Malo.

Here the troops were landed on the 4th and 5th, though not without loss, as several were drowned. Morlaix was the place next designed to be attacked; and a storming of St. Malo from the French or land side was projected, but this was found to be absolutely impracticable. Meanwhile the elements were, as usual, against the expedition. It became impossible for the fleet to remain where it was; certainly it would have been impossible to re-embark the troops there. Commodore Howe (now Lord Howe, owing to his brother's death) found that the only available anchorage near was the bay of St. Cas, and proceeded there. Accordingly the army marched, and reached St. Gildan on 9 September and Matignon on the 10th. Here they came near the fleet. That evening news came that twelve battalions of foot and two squadrons of horse were on their way from Brest to attack the British. Bligh then retired on St. Cas and sent off an officer to acquaint Lord Howe, in order that all might be prepared for the embarkation. The British reached St. Cas Bay at 4 a.m., 11 September. Here the flat-bottomed boats were found ready drawn up on the shore. The troops immediately embarked. About an hour after the enemy appeared and opened artillery fire from the heights above the shore, but did not descend until all the troops except the Grenadiers were safely embarked.

These, forming as they did the rear guard, advanced to meet the enemy, and behaved with great bravery, till, being outnumbered and overpowered, they were compelled to retire to the waterside and defend themselves as best they might till the arrival of the boats to take them There was considerable loss, between 600 and 700 men being killed, drowned, wounded, or prisoners. General Drury was killed, Lord Frederick Cavendish taken prisoner, and at least ten officers were missing. Lieut.-Colonel Wilkinson was killed. After this disastrous end to a practically useless campaign the fleet and transports made sail for England, and reached its shores about 18 September. Bligh had despatched his captured guns home directly after the taking of Cherbourg.

On 16 September we read that these trophies 'passed by his Majesty,' and set out from Hyde Park, through the City, and proceeded to the Tower 'in a grand procession, guarded by a company of matrosses' (artillery men} with drums beating and fifes playing all the way.

The guns, we read, were 'finely ornamented with the arms of France ami other hieroglyphics, such as trophies, etc.' All the guns except six, we are told, 'remained spiked as they had been by the enemy before abandoning them.'

Each gun, it may also be added, had its name, exact weight, etc., inscribed on it or cast with it. The names were some of the classical, such as Hecube, Nitocris, Antonin; among others were Téméraire, Insensible, Violente, Foudroyant, Impérieuse, and Diligence.

Chapter 10

Battle of Warburg

1760

As has been already stated, the 7th Queen's Own Dragoons were warned for foreign service towards the end of March 1760. Accordingly, the Heavy Troops, six in number, embarked in the river Thames early in April, their destination being Germany. The Light Troop, it may be remarked, remained at home, being stationed at Hitchin. For a wonder the Regiment had a prosperous voyage and a quiet passage to the river Weser, where, under the command of their Lieut-Colonel, George Lawson Hall, they disembarked not far from Bremen in Lower Saxony. At that time the British force on the Continent was under the command of the Marquis of Granby. Of the Allied Army, which included Hanoverians, Hessians, and Brunswickers, Prince Ferdinand of Brunswick was in chief command. His camp was situated on the heights of Fritzlar in Lower Hesse, and thither the 7th Queen's Own Dragoons proceeded, and reached their destination on 21 April. On arrival they were brigaded with the dragoons of Priceschenic, their brigadier being a Colonel Bremar.

On 5 May the Allied Army advanced into Hesse.

A detachment of the army was left in the Bishopric of Minister under General Spörcken to watch the proceedings of the French Army of Reserve which, under the command of the Count de St. Germain, was approaching from the Rhine.

Prince Ferdinand had garrisoned four towns, Cassel, Dillenburg, Marburg, and Ziegenhagen. Near Kirchain on the river Ohm the passes of that river were held by General Imhoff against the advance of the main French Army, which under the command of De Broglie was approaching from Frankfort.

Meanwhile, Frederick the Great had recalled the detachment of Prussian cavalry which had been attached to the main Allied Army. Ferdinand, with whom was Lord Granby, was now on the march for the heights of Hombourg near the river Ohm. De Broglie advanced. Imhoff, whether owing to imperfect instructions or to some other cause, did not oppose the passage of the Ohm, and De Broglie safely crossed, to the mortification of Ferdinand. Imhoff shortly afterwards resigned in consequence of the scathing condemnation of his supineness passed on him by his chief.

The passage of the Ohm by De Broglie had these results. He obtained the command of the river, the possession of Amöneburg and Marburg, and was also enabled to occupy the heights of Hombourg long before Ferdinand could possibly reach his goal On the news of this check Ferdinand halted at Ziegenhagen and then retreated to Sachsenhausen, where he encamped. Hitherto the British cavalry had been in the first line of the Allied Army, but the ground being found most unfavourable for cavalry action, a change was made and they were transferred to the second line. This change was the outcome of a Council of War held on 27 June.

Meanwhile De Broglie was hastening his advance. He passed through Neustadt, Rosenthal, and Frankenberg, and arrived almost at the heights of Corbach, a position at that time held by General Lückner. The opposing forces were now separated by no more than a three hours' march. A series of small skirmishes took place, in one of which near Zielbach the Hereditary Prince nearly lost his life.

The Comte de St. Germain was now approaching with the French Army of Reserve, and the nearer he got to Sachsenhausen the more hazardous did the situation of Prince Ferdinand and the Allied Army become. True, his position was a strong one, but he was in constant danger of being out-flanked, and out-flanked too by vastly superior forces.

The Hereditary Prince was therefore on 10 July despatched with a mixed German and English force to Corbach, which St. Germain had already succeeded in occupying after driving out the meagre force under General Lückner which had been posted there for its defence.

Here the Prince met with disaster, though, owing to a gallant cavalry charge, his infantry was enabled to retire in order. The whole of his right brigade of artillery was captured. A force of the enemy at the same time threatened the outposts o the camp of Sachsenhausen, thus preventing Granby from giving any assistance to the hardly-pressed troops at Corbach. This unfortunate affair cost the Allied Army 800 killed and wounded and eighteen guns. The Hereditary Prince, however, amply revenged himself for this defeat at Emsdorf on 14 July.

De Broglie and St. Germain now effected a junction at Corbach. where, finding that though senior he was to act in a subordinate command to the former, St. Germain retired, being succeeded in the command of the French Army of Reserve by the Chevalier de Muy. De Muy was now sent across the river Dymel and encamped with between 25,000 and 30,000 men, with his right on the town of Warburg and his left on the heights of Ossendorf. De Broglie remained at Corbach, where he occupied his time in harassing both the flanks of the Allied Army. The position taken up by De Muy too at Warburg threatened to cut Ferdinand's line of communications with both Westphalia and Hanover. There appeared no other course open but to retreat, and accordingly on the night of 24 July this took place, the destination of the Allies being Cassel. The retreat was carried out in perfect order and without any molestation on the part of the enemy. Lord Granby held the command of the rear-guard, having with him four Major-Generals, Schlüter, Honywood, Elliot, and Griffin.

Passing over the heights of Freienbergen the retreating army reached the plain of Kalle, which is situated about ten miles to the north-west of Cassel. Here they encamped, and then continued the retreat to Wilhelmsthal.

It will be readily understood that these continuous marchings involved much fatigue to the Allied Army; nor was this all, for the enemy, which largely outnumbered them, held positions which threatened them on all sides. What Prince Ferdinand desired to do was this, to keep the river Dymel open, as without it his communications with Minister, Paderborn and Osnabrück would be cut. Cassel was in his rear and Hesse required to be covered. How these objects were to be effected was thus. While apparently in force

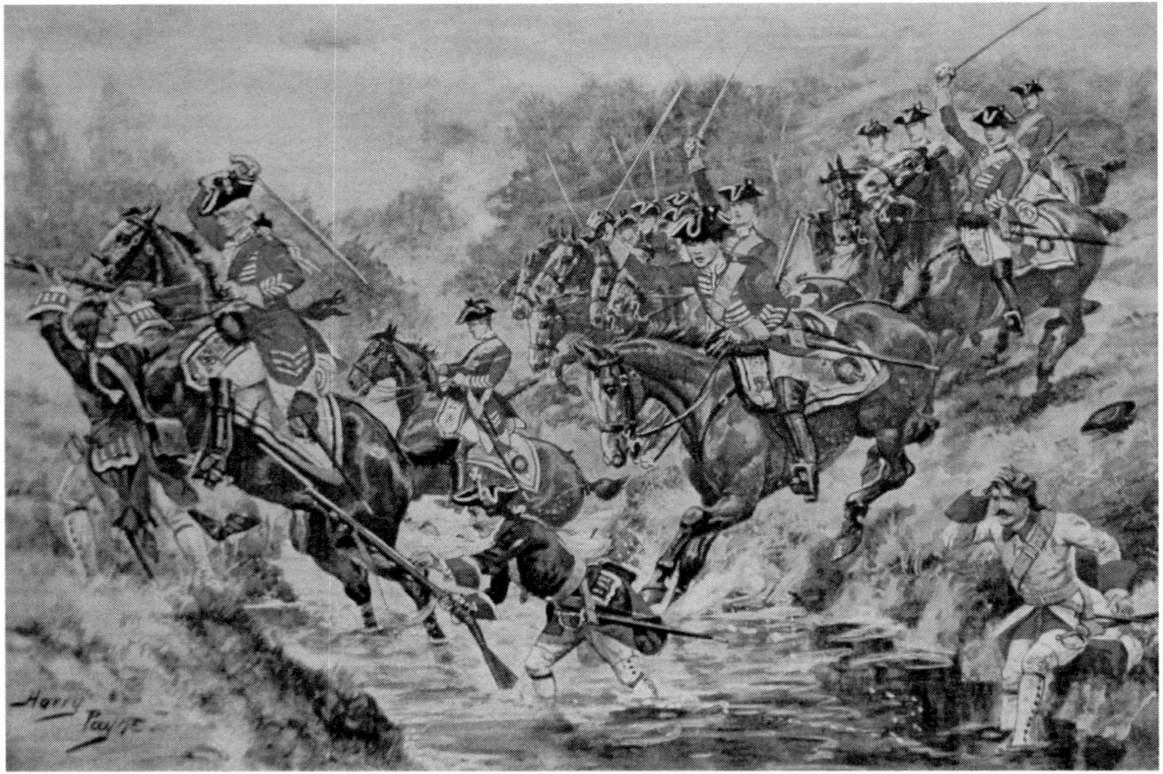

Warburg-1760

before Cassel, he meditated a stroke by which the Chevalier de Muy would be cut off on the river Dymel. Meanwhile De Broglie threatened the right of the Allies and Prince Xavier (the Comte de Lusace) threatened their left. If Ferdinand crossed the Dymel, Cassel must inevitably fall. Ferdinand, however, determined to proceed on his expedition.

Leaving Count Kilmansegge and General Lückner to cover Cassel, but with orders to retire on Münden if they should be attacked by superior forces, he ordered the Hanoverian corps under General Spörcken to proceed to Liebenau on the river Dymel. Following Spörcken he sent a force under the Hereditary Prince. This force was composed of the British Legion (a composite body mainly consisting of foreign auxiliaries), two battalions of grenadiers, made up of the grenadier companies from various regiments, the Highlanders, two squadrons of the 7th (Queen's Own) Dragoons and two of Conway's Dragoons.

The movement began on the morning of 29 July, 1760, when the troops crossed the Dymel and took up a position between Liebenau and Korbeke. Thanks to a dense fog the movement was carried out successfully.

Ferdinand himself left the camp at Kalle at an hour before midnight, on 30 July, and crossed the Dymel. At 5 a.m. on 31 July the entire force was drawn up on the heights overlooking Korbeke. Lord Granby commanded the right wing, which consisted of the British cavalry and three brigades of artillery. De Muy lay at Ossendorf.

The Hereditary Prince and Spörcken were now despatched towards Dossel with orders to turn the left of De Muy at Ossendorf. Again a heavy fog enabled this movement to be successfully carried out, and the Allied force was permitted to gain a valley and to advance unperceived by De Muy till it suddenly appeared at 11 in the morning. Ferdinand and Granby followed them with the main army. The British cavalry were now posted in the second line and were commanded by General Mostyn.

The attack on De Muy began and began successfully. His left was driven in on his centre and right. Time, however, was wanting to permit the infantry to arrive and, delayed by bad roads, they were still some distance from Warburg—some accounts say as much as five miles—when the action began.

Ferdinand remained with the infantry, but detached Granby with the British cavalry and artillery to press on in advance. At a trot the whole of the British cavalry advanced in order to accomplish the five miles which separated them from the town of Warburg. They duly reached Meine and formed up. Their appearance immediately caused the enemy to waver. The details of the battle are as follows:

The Hereditary Prince and Spörcken advanced in two columns—one, which included cavalry, artillery, and infantry, composed of the Royal Dragoons and the grenadier battalions, headed for Gross Eider and Ossendorf; the other, headed by the 7th Dragoons with the Highlanders of Keith and Campbell, followed them, acting as cover to the grenadiers in the second line. The artillery of the Hereditary Prince, posted on the outskirts of Poppenheim opened fire at 1.30 p.m., and at the same time the grenadier battalions pressed through Ossendorf, certain French troops placed there to protect the left wing of De Muy retiring before then without offering any resistance. The Allies then advanced towards a steep hill in the rear of the enemy, and were evidently intent on seizing it; thereupon a French battalion belonging to the Regiment Bourbonnois faced about and returned, with the object of forestalling the Allies in the possession of this important position. It was in the nature of a race as to which party could first secure a lodgement there. Ten grenadiers headed by Colonel Beckwith advanced at a double, followed by the Hereditary Prince himself leading thirty more, and breathless they arrived first at the crest. On came the battalion of the Bourbonnois and were received by a galling fire. Ignorant of how strongly the hill might be held, the enemy halted and awaited the arrival of their second battalion, which had been sent to support them. A delay of about ten minutes thus took place, and this was a sufficient time to allow the remainder of the grenadier battalion to come up and to join the little band at the top of the hill.

Then the struggle began, a struggle be it observed which was maintained by one British battalion against two of the enemy. The fight was most strenuous and undoubtedly in the end the disparity of numbers most have told against the British had not a battalion under Maxwell arrived to their assistance. Again the fight was resumed, the enemy stubbornly contesting for the possession of the coveted hill. They were, however, unable to prevail, and additional troops were in the act of being sent to their assistance. Nor was assistance withheld from the gallant band of British, who still, though with difficulty, maintained their hold on the hill. A battery of artillery was despatched to their aid, but in the passage of a defile near Ossendorf this in some way broke down and blocked the road, thus preventing the advance of the remainder of the column. For a brief time it looked as if the grenadiers would be overwhelmed, but by good luck or hard work the passage was cleared, and the remainder of the column gained the hill and entered into the fray. Almost simultaneously the advanced guard of the other column arrived, and taking the reinforcements which the enemy were hurrying up towards the hill in flank, put them into disorder. A charge of cavalry, in which the Royal Dragoons and the 7th Queen's Own Dragoons took part, completed the discomfiture of the already shaken enemy.

The cavalry in this operation slew many and captured a goodly number of prisoners. The upshot of this was that the enemy's flank was turned, though whether De Muy could retrieve this reverse remained to be seen. Up to this period it must be remembered that the first column only had been engaged and the other column had not yet come into action. As a matter of fact it had not been able to arrive on the field as a whole. A portion was, however, in position and was threatening the left of De Muy. The Allied main body was still some miles away, struggling vainly to cover ground which was marshy and very unfavourable to troops already wearied by a long and toilsome march. It is stated that many fell by the way from sheer fatigue, for the weather was extremely hot. And then it was that Prince Ferdinand sent forward Lord Granby with the British cavalry and artillery, as has been already stated. This advance took place just in the nick of time, as they arrived on the field at the moment when De Muy was about to attempt to retrieve his reverse; also too, Granby's command appeared in the front of the enemy's line, and appeared still advancing at a trot. It is stated that the pace of the guns on this two hours' advance was a matter of astonishment to those who beheld it.

Immediately on arrival Granby halted momentarily to form his cavalry into two lines, and then his twenty-one squadrons hurled themselves on the cavalry of the right wing of De Muy. Hatless and with his bare bald head Granby rode at the head of the Horse Guards Blue, and straight at the enemy.

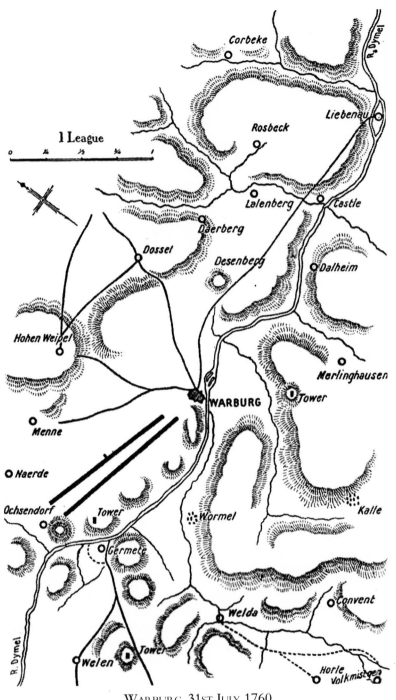

WARBURG, 31ST JULY 1760

Only three French squadrons had the hardihood to await the onset; the remainder turned tail and made off. Granby's squadrons therefore wheeled and turned their attention to the French infantry, whom they took in flank and rear. The three French squadrons before mentioned, however, behaved with great gallantry, and charging on the flank of a British cavalry regiment handled it very roughly, only in their turn to be cut to pieces by the Blues.

The fate of the French cavalry had its effect on the morale of their infantry, and it resisted not, nor indeed is that a matter of much marvel, for it was open to attack on both flanks and also in the rear. A disorderly the flight ensued; following the fleeing cavalry, and indeed mixed with any them, the entire French Army made for the river Dymel, into which they plunged, at fords if they could find them, elsewhere if they could not.

Arms were thrown away and a complete *sauve-qui-peut* ensued. A body of irregular French troops in the town of Warburg itself attempted to escape, but was caught by the pursuing cavalry and perished nearly to a man. To keep the retreating enemy on the run ten squadrons cavalry were despatched across the river after the British ad been brought down to the bank had done considerable execution on the fugitive French. As a victory, Warburg was absolute and complete. The fugitives continued their disorderly flight till they drew breath at Volksmissen, a distance of ten miles from the scene of the battle.

The losses of the Allies on this occasion amounted to about 1200 killed and wounded, a large proportion of which belonged to the grenadier battalions and fell while disputing the possession of the hill.

The cavalry suffered but little, except in the case of the regiment which was engaged with the three French squadrons previously noted. The casualties on the French side amounted to from six to eight thousand men killed, wounded, and prisoners; they also lost twelve guns, is the estimate of Mr. Fortescue. In the Life of the Marquis of Granby by Walter Evelyn Manners we read 'The Chevalier de Muy left 1500 on the field, besides 2000 prisoners, 10 pieces of cannon and his papers.' The latter seems in all probability to be a more accurate statement of casualties, and we prefer to accept it in preference to the other. Cannon does not give us much information on this battle; he writes:

The cavalry under the Marquis of Granby and Lieut.-General Mostyn arrived at a favourable moment; a gallant charge of the British squadrons decided the fortune of the day, and the French made a precipitate retreat across the Dymel. The Seventh Dragoons supported the infantry in the attack on the enemy's flank, and by a spirited charge towards the close of the action contributed to the success of the day. The conduct of the British Cavalry was commended by the Marquis of Granby in his public despatch, and Prince Ferdinand declared in General Orders, that "all the British Cavalry performed prodigies of valour." The Regiment being eager in the pursuit, had four men and horses captured by the enemy; three of the men, however, escaped and rejoined the regiment.

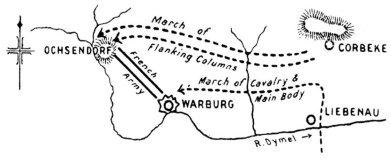

WARBURG, 31ST JULY 1760

The Manuscript Regimental Record is even more niggardly in the information it vouchsafes.

> 31st July. The Regiment was present at the Battle of Warburg. Extract from the returns of Killed and Wounded inserted in the London Gazette.
>
> > Cope's Dragoons. Killed, none—Wounded, none. One private and four horses missing.

In the Additional Hardwicke M. S. S. No. 35839, folio 203, will be found Prince Ferdinand's narrative of the Battle of Warburg. The account is in French, and fills fourteen quarto sheets of paper closely written. We propose here not to give it in its entirety but to quote from it considerable extracts, not in the original language, but in English. He states that the battle was fought between the. . . .

190

.... reserve of Chevalier du Muy, twenty-eight battalions and thirty-eight squadrons, and the reserve of M. de Spörcken, joined by a corps under the command of the Hereditary Prince of Brunswick, in all twenty-four battalions and twenty-two squadrons, sustained by twenty-two squadrons of the right wing under the Marquis of Granby.

The account begins thus:

Marshal Broglie manoeuvred to cut them off from Westphalia, and to force the King's Army to fight at a disadvantage in the cul-de-sac formed by the junction of the Dymel and the Weser; or to pass the river to assure himself of the possessions of Hesse and to cam- the war into Hanover.

The Allies took fitting measures to upset this design. Liebenau on the Dymel was occupied by the British Legion on 26 July. On the same day Major-General Scheiter occupied the heights opposite Liebenau on the right bank, with three battalions of grenadiers to support the British Legion.

Learnt on the 28th that Du Muy was advancing on Warburg with the reserve—on which M. de Spörcken, who was encamped with fourteen battalions and fourteen squadrons on the right of the army at Westuffeln, was ordered to pass the Dymel on the 29th and to camp between Liebenau and Korbeke. The Hereditary Prince of Brunswick joined him there on the night of the 29th or 30th with ten battalions of Grenadiers and Montagnards and eight squadrons, half German and half British.

Lord Granby replaced General Spörcken with six battalions and four squadrons at Westuffeln, and the Prince of Anhalt occupied the camp of the Hereditary Prince of Heckewhausen, whence he detached three battalions of Grenadiers to Obervolmar to join Major-General Wolfe, who was there with four squadrons. News came on the 29th at night that du Muy was about to be reinforced. At daybreak on the 30th the enemy's advanced guards debouched at the same time in several places towards the army camped at Kalle and towards the corps posted on the flanks. This forced Major-General de Lückner *de se replier* with the light troops under his orders at Niederzweren sur Cassel. Winter Kasten was occupied by the reserve of Mons. le Comte de Lusace, who opened fire at once on Count Kilmansegge, who was encamped in the entrenchment at Cassel. In the afternoon the columns debouched by the Duerenberg and Zierenberg towards Weimar and Furstenwalde.

Shortly after this Wolfe's outposts at Weimar were driven in and the Duke sent supports from the 2nd Line to sustain them, and all the cavalry of the left wing marched out to support the Prince of Anhalt at Heckershausen. Several hundred cannon shots were exchanged, after which the troops returned to camp. The Duke no longer doubted that the enemy's whole force had either arrived or was on the point of reaching the environs of Cassel, which was a long march distant from Warburg. He resolved therefore without further delay to pass the Dymel and to fight Du Muy. On July 30 the army marched in eight columns at 9 p.m. and passed the Dymel before the enemy perceived what was in the wind. The Prince had reconnoitred Du Muy's position on the 30th; his right rested on Warburg, his left extended in front of the village of Ocksendorf, a camp large enough for "20/m" men. They had discovered a little camp on the right bank of the river Dymel in the Wood of Welda, but could not quite determine its position. This camp disappeared (was struck) in the afternoon, and news came that the troops there had crossed the Dymel and joined Du Muy, who had not altered his position.

The Prince then made his dispositions for the attack and sent them to the Duke, who approved. Spörcken's reserve and that of the Hereditary Prince joined together made up twenty-four battalions and twenty-two squadrons, but the battalions had been much weakened by detachments which had been sent out, and only numbered in some cases fourteen effective men.

They marched in two columns; on the right the infantry of the right. A brigade of heavy artillery was brought by Spörcken and passed by Borgenbuk, having Eissen on the right, Grossen-Eder on the left, and going through Necdern and Ocksendorf. The column was ordered to pass the tower which was in the rear of the enemy's left and to form on the two lines of foot and on a third line of cavalry. Lieut.-General de Zastrow brought the column from the left of the first and second line of the cavalry of the left and of two brigades of artillery. This column was ordered to pass by Coerbeke, Dinckelbourg, Kleinen-Eider, and Menne, to deploy in two lines of foot and a third of cavalry, with the left resting on Menne and the right extending to Ocksendorf. The Prince attacked on flank and rear with the reserve while the army itself advanced towards the front with its right on the Menne and its left behind the town of Warburg, against which place the British Legion was to direct a false attack.

The heads of the columns debouched about 6 a.m. on the heights of Coerbeke, the passage of the Dymel having delayed them a little. They had hurried, but with all the will in the world to go quickly could not manage to arrive as early as had been intended. But in order to lose no time Spörcken's troops started at 7 a.m. and marched to their destination to turn the enemy's flank. M de Bülow *qui avait été poussé la veille* with the British Legion at Desenberg was attacked there at daybreak. He retreated to the village of Rosbeck, where he received intelligence that the enemy was moving towards him.

A thick fog which concealed us from the enemy also prevented our seeing what he was about. They were not troubled on the march except by a brook which had to be crossed and a marsh which had to be avoided. Still these obstacles caused delay. At 1.30 the Hereditary Prince began the attack with the reserve of Spörcken. Colonel Huth commanded the established a battery of four guns (twelve-pounders) on the right of the village of Menne, another of equal strength on the left of the village of Ochsendorf, to cover the debouchement by the village. A third battery was posted *en deça* of the village of Ocksendorf behind *une butte de terre, tout attenuante au village.*

The two first pegged into the flank of the enemy and the last took them in the rear. Everything was going *à point nommé*, and the batteries opened fire the moment the head of the column entered Ocksendorf. The enemy's troops posted there replied on our arrival *sans lacher leurs coups*. They had some battalions posted on their left who were drawn up *en potence* and when they saw that we were bringing up a considerable force on a *hauteur escarpée* which was on their rear they marched the Regiment de Bourbonnois to hinder them. Colonel Beckwith commanding the English Grenadiers, who were at the head of the column with Mr. Wargot the Aide-Major (Adjutant) of the Brigade, sent in advance une dixaine of Grenadiers, who brought the news to the Hereditary Prince that the enemy was pressing towards the height very briskly. On which the Prince himself pressed forward with a *peleton de* 30 Grenadiers on to the height to hinder them.

The French, who were on the slope of the hill, could not see whether this *peleton* was supported or not, and halted for five or six minutes. This gave the battalion of the Grenadiers de Daulhal time to come up. The fire then got very hot, for at this moment the 2nd Battalion of the Bourbonnois Regiment joined the first. De Daulhal's men were then pressed hard and began to give way, when Maxwell arrived at the head of the 2nd Battalion of English Grenadiers, who came from the other side of the mountain, leaving the tower on their left. They gave time for de Daulhal's men to re-form and advance again, and this they did in good style, thanks to their officers and to the entendement and activity of Beckwith. Between the British Grenadiers and the Hanoverian Grenadiers a train of ten twelve-pounders came on. But in passing an *assez mauvais* defile by Ocksendorf, nearly half an hour had been lost before they cleared the village and the other battalion could join.

They got up, however, at the most critical moment, just as the enemy was detaching other regiments to reinforce the Bourbonnois. The Swiss Regiments of Jenner and Lockmann formed this reinforcement and were charged by the 4th Hessian Guards, who headed the left column which was commanded by de Zastrow, who debouched precisely on the left flank of the enemy. The battalions got into action on arrival, *s'empara tout à fait de la hauteur de la tour*, and chased the enemy from one height to the other. The regiments of

Conway and Cope's Dragoons (7th Hussars), profiting by this, fell on the fugitive infantry and made many prisoners. The Duke perceiving that the difficulties of the road would prevent the infantry, wearied as they were already, from getting ahead quickly enough to arrive in time, ordered Lord Granby to advance with the whole of the cavalry of the right: twenty-two squadrons of British. Count Schaumberg-Lippe was ordered to advance at the same time with the British Artillery. This he did with admirable promptitude.

Lord Granby led the charge on the enemy's cavalry which was opposed to him. It wavered without awaiting the shock, with the exception of three squadrons which took Bland's Regiment in flank, at the very moment that the British Cavalry was taking the enemy's infantry in flank and rear—that is to say the infantry engaged with the men of the Hereditary Prince. Bland's Regiment was *dégagée* on the field by one or two squadrons of the Blues, commanded by Lieut.-Colonel Johnson, and the enemy's cavalry was *maltraité au possible*.

The British Cavalry in its advance *tint l'ennemi en suspense,* and favoured by that the formation of the column of M. de Zastrow. It fell on with such good will that the enemy, finding themselves pressed on all sides, at once gave up the combat and retired in great confusion. The greater part both of infantry and cavalry rushed headlong into the Dymel, not crossing by a bridge but *à qué*, many throwing away their arms *pour v'alleger*. M. de Dulow, having attacked Warburg town with the British Legion, *en deburgna le corps de Fischer qui tombant ensuite* beneath the swords of the British Cavalry, was nearly annihilated. Count de la Lippe played on the enemy with the British Artillery during the retreat with great success and pursued them along the heights of the Dymel, and when they tried to re-form on a height near a wood not far from Welda, on the right bank of the Dymel.

He kept them running, however, when once they had started. The Duke and Lord Granby now passed the Dymel with ten squadrons and twelve British battalions. The enemy retired on Volksmissen and Lord Granby occupied the high ground of Welda. The Hereditary Prince ramped on the field of battle with Spörcken's victorious reserve. The army camped between Menne and Devenberg, the Prince of Anhalt and de Wagenheim at Liebenau and Lammeus. Eight officers and one hundred and ninety-four non-commissioned officers and men were killed, fifty-five officers and nine hundred and two non-commissioned officers and men wounded, one officer and seventy-seven men missing; so that our entire loss was but one thousand two hundred and thirty-seven men. One cannot too highly praise the willingness and determined resolution of the victors as displayed on this occasion. All did well, without excepting one single battalion or one single squadron. Fortune perhaps most favoured Maxwell's battalion, that of Bock, the Hessian Grenadiers, the 4th Hessian Guards, and all the British Cavalry. The artillery did what could not have been better from beginning to end: their shooting was most accurate. Twelve guns and twenty-eight ammunition wagons were taken.

It is difficult to assess the loss in men (of the enemy), but it must be almost one thousand five-hundred killed and as many wounded, besides those drowned in the Dymel. There were also many prisoners taken.

We think that our readers will agree with us that despite its length the above document is well worth its inclusion in these pages. Certainly among the accounts of the Battle of Warburg it is decidedly the best.

There is not one entry for the year 1761. The result of this brilliant victory was, however, discounted by the loss of Cassel and all the immense collection of valuable military stores which it contained. Prince Xavier advanced against Kilmansegge with a greatly superior force and compelled him to retire, first on Münden and subsequently across the river Weser. Cassel then fell. A detachment of French troops next laid siege to Ziegenhagen under the command of the Comte de Stainville. The siege lasted six days, after which its German garrison surrendered as prisoners of war. Here again there is a divergence of statement between Fortescue and Manners, as the former tells us that this siege lasted for ten days.

After the battle of Warburg, Prince Ferdinand remained quiescent on the Dymel, and here he was menaced by the main French

Army under Broglie, who moved in an easterly direction towards Immenhausen. This carried the war uncomfortably near to the frontier of Brunswick. Ferdinand, however, being compelled to preserve the fortress of Lippstadt, could not move to the east bank of the river Weser without danger to the safety of that place, as its preservation prevented concerted action between the two French Armies, those of the Rhine and the Main.

Ferdinand now resolved on a new enterprise, as, finding the garrison of French troops at Wesel had been weakened, he determined to attack it. Making preparations therefore which consumed the earlier portion of September, he despatched a heavy train of siege artillery to Wesel and followed it up three days later by a force of 10,000 Hanoverians and Hessians under the Hereditary Prince. Broglie sent a force in pursuit of him under the Marquis de Castries. Ferdinand next despatched a reinforcement of British and Hanoverians to the Hereditary Prince consisting of both cavalry and infantry, but among the former the 7th Queen's Own were not included.

The Hereditary Prince laid siege to Wesel. Castries by a splendid march followed in hot pursuit, and en route was joined by reinforcements from Brabant. This was an awkward move for the Hereditary Prince, as, when allowances were made for the besieging force detained before Wesel, he found himself with numerically a very inferior force to oppose his now augmented enemy.

He must therefore either fight an unequal battle or raise the siege and utilise his entire force. He chose the former alternative. The battle of Kloster Kampen followed, but the story of it is in no way concerned in this history. After the battle the Prince retreated northward to Büderich, where the bridge having been destroyed by floods the river was impassable. His retreat was cut off and his ammunition was nearly if not quite exhausted.

However, he entrenched himself, in part using his wagons for that purpose, and repaired the bridge.

A crossing was effected on 18 October, and a junction with the besiegers of Wesel took place. The retreat continued, during which the Allies were energetically pursued by Castries. Eventually the Prince took up his quarters in Westphalia in order to protect Lippstadt and Münster. Thence he detached a force to assist Ferdinand, who was by this time in almost as dangerous a situation as himself.

The end of the campaign was now at hand. Ferdinand made an unsuccessful attempt to obtain possession of Göttingen and then went into winter quarters at Warburg.

The situation was now this. The French held Hesse, Göttingen, and the defiles near Münden. thus having the path to Hanover and Brunswick open to them, and their headquarters were fixed at Cassel. The 7th Queen's Own Dragoons were encamped for the winter in huts specially built for them near the banks of the Dymel; by which means they and their horses were protected from the severity of the weather. Later they went into cantonments in the villages in that part of the Bishopric of Paderborn.

Regimentally an important event took place in this year.

On 28 July the Colonel, Lieut .-General Sir John Cope, K. B., died, and was succeeded in the Colonelcy by Lieut.-General John Mostyn, who was transferred from the 5th Royal Irish Dragoons by the special wish of George II, and as a reward for his gallantry at Warburg, of which the mention in despatches had been most eulogistic. This transfer was somewhat out of the regular course, as the 5th was a royal regiment, was senior in the service, and was of a stronger establishment. Mostyn's personal view of his promotion is treated in Appendix 3.

It is somewhat curious to note that by a vulgar error the 7th Queen's Own Dragoons, because their Colonel was Sir John Cope, have often been confused with Gardiner's Dragoons, who suffered defeat at Preston Pans, where Sir John was in supreme command. It is, however, hardly necessary to add that the 7th Queen's Own were not present at that unfortunate battle.

CHAPTER 11

Battle of Vellinghausen
1761

We left the 7th Queen's Own Dragoons cantoned in the villages of the bishopric of Paderborn near to the River Dymel. In this position they remained until February, early in which month Prince Ferdinand determined on the initiation of a winter campaign, hoping thereby to retrieve the unfavourable issue of the campaign of the previous year. On the 11th of the month he put his troops in motion, trusting by boldness to succeed in recapturing Hesse, the whole of which State at the close of the preceding year remained in the possession of the enemy.

The Allies marched in three columns. Spörcken, who commanded the eastern column, was despatched to the Werra and Unstrut to effect a junction with the Prussians and then to fall upon the Saxons. The Prince himself led the central column towards the Eder. The third or western column from Westphalia, and under the command of the Hereditary Prince, was ordered to advance against Fritzlar, while a separate force was detached with the view of gaining possession of Marburg.

Of these three projects only that of Spörcken succeeded, and that general gained a notable victory at Langensalza. Ferdinand and the Hereditary Prince both failed. True, Prince Ferdinand managed to drive the enemy out of Hesse, but he was unable to hold the territory, seeing that neither Cassel, Ziegenhagen, or Marburg had yet been reduced. The country had been devastated, the roads were impassable, and heavy falls of snow added to the hardships which the army had to endure. There had also been a great mortality among the horses owing both to the lack of forage and also to the inclemency of the weather. Ferdinand then attempted by marching southwards to gain some more hospitable region, and succeeded in reaching the Ohm. De Broglie, however, advanced northwards to meet him, and having arrived at Giessen it became advisable for Ferdinand to retire. He therefore fell back to the Eder on 20 March.

The enemy on the next day attacked the Hereditary Prince at Grünberg, and by sheer superiority in numbers forced him to retreat, during which he lost about 2000 prisoners.

This serious check to one of his columns compelled Ferdinand to raise the siege of Cassel and to retire with as much speed as possible, having De Broglie not only in hot pursuit, but possessed moreover of a force of numerically almost four times the strength of his own. Towards the end of March Ferdinand had regained his old quarters to the north of the river Dymel, and here he again sent his wearied troops to recuperate in cantonments.

The losses in men, horses, and stores in this campaign were extremely heavy on both sides, and a few weeks' respite from active hostilities was as welcome to the enemy as it was to the Allies.

The time thus employed was not wasted by France, for early in April the Prince de Soubise was despatched to the Lower Rhine and instructed to co-operate with De Broglie; his army now numbering 100,000 men. De Broglie, on the Main, commanded 60,000 troops. The task of Soubise was to advance against Ferdinand and force him to abandon Westphalia, Münster, and Lippstadt, while the French force in general rested and recovered from the toils of the winter campaign. If Ferdinand did not withdraw he must fight, and in either event he would be compelled to keep his men continuously on the move, while those of De Broglie remained quiescent and prepared for a projected invasion either in the direction of Hameln on the Weser or into the electorate of Hanover. Such a course, if pursued by De Broglie, would have necessitated the immediate opposition of Ferdinand, in which case the large army under Soubise would have been enabled unopposed to act in any direction its commander pleased. It must, however, he remarked that owing to Court intrigues the larger force had been given to Soubise, and also that no orders as to genuine co-operation in the campaign had been given. Consequently, the two French armies were acting independently. The situation was therefore this. Soubise, the inferior soldier of the two, commanded the largest force; while De Broglie, by far the more skilled commander, had at his disposal an army of not more than three-fifths of that directed by his rival. Ferdinand had a force which, while it was slightly inferior to that of Soubise, was numerically superior to that of De Broglie, for his numbers amounted to about 93,000 men.

Obviously speedy action on the part of the French was the proper course to pursue, but this course was not followed. True, Soubise reached Frankfort by 13 April, where he had a meeting with De Broglie, whom he had summoned for that purpose. What subjects were discussed, or what plans (if any) were formed on the occasion, is not known. One thing, however, is certain, that for the remainder of the month, for the whole of the next and until the beginning of June Soubise did nothing. When indeed he did make a move he selected as the field of his operations the country south of the river Lippe and between that river and the river Ruhr; his intention being apparently to join forces with De Broglie. Possibly, too, he felt considerable unwillingness to engage in a pitched battle with Ferdinand, for whose skill as a commander and for the fighting powers of whose army he had a wholesome respect.

Ferdinand on his part first detached a force under the Hereditary Prince to watch Soubise from a position a little to the westward of Minister. Meanwhile he toiled assiduously at the tremendous task of preparing both his army and his transport for the approaching campaign. This occupied the Prince for about ten weeks.

Soubise then advanced and crossed the Rhine on 13 June, penetrated as far as a place called Unna to the east of Dortmund by the 23rd, and there he entrenched himself.

Ferdinand, who had concentrated his army four days previously at Paderborn, advanced to the west to meet Soubise, but took the precaution to leave a small force of observation before Göttingen, and a corps of 20,000 men under Spörcken on the river Dymel to watch De Broglie.

Ferdinand marched on the 20th, and eight days afterwards fronted Soubise and pitched his camp. The Hereditary Prince, who had been also advancing, here effected a juncture with Ferdinand. Inspection of the French position showed that it was apparently too

H. S. H FERDINAND
DUKE OF BRUNSWICK

THE MARQUIS OF GRANBY

MARSHAL DE BROGLIE

strong to be attacked with any reasonable success, and Ferdinand then determined upon a somewhat curious manoeuvre. This was by means of a forced march round the left flank of Soubise to drive that commander into a junction with De Broglie. This march occupied thirty hours, and was made by way of Camen. Suddenly Ferdinand's army appeared full in the rear of Soubise at Dortmund. From this position he attacked that commander on 4 July, and without resistance from the enemy succeeded in his object, for Soubise, with the Allies in hot pursuit, made off for Soest, whither De Broglie at once proceeded for a second consultation. Meanwhile Spörcken had been driven away from Warburg and the river Dymel by De Broglie, and in the operation had lost some guns.

De Broglie had then turned westwards, had occupied Paderborn, and it was from Paderborn that he proceeded to Soest. Here he was followed almost immediately by his army, and the joint commanders had then at their disposal 100,000 men. This largely outnumbered the troops of the Allies, who, even with Spörcken's corps of observation, winch had rejoined Ferdinand from its post on the Dymel, could not muster more than 60,000 available fighting men. The odds were therefore heavily against the Allies; but Ferdinand notwithstanding maintained his ground. It was open to him to have crossed the river Lippe, but this would have delivered Lippstadt to the enemy. Undoubtedly the French desired him to take this course, and his refusal so to do was wise. Ferdinand therefore elected to fight, and in consequence manoeuvred his army for some days until on 11 July he took up a definite position.

His main army remained encamped on the south bank of the Lippe. Spörcken with 8000 men was detached across the river at Herzfeldt to observe the movements of Prince Xavier of Saxony, who had been Ferdinand's left wing was extended to Kirchdenkern, a village on the Ase, a stream without fords and only passable by means of bridges, which were few in number. His headquarters were at Vellinghausen, a spot half-way between the Ase and the Lippe and beneath a hill called the Dünckerberg.

We will now return to Lord Granby, who had left the Continent for England when the Allied Army first went into winter quarters.

He left England to return to his command on 1 June, and finding that travelling by the direct route from the Hague via the Wesel was not safe, was compelled to take a circuitous path thither, by way of Zwölle and Osnabrück, thus only arriving at Paderborn on 13 June. Here he found the British cavalry in good condition, though the number of sick in the entire army was large, and is stated to have amounted to 12,000 men.

Immediately on his arrival he was able to report that the British cavalry and the Brigade of Guards would be leaving their winter quarters for a camp near Paderborn, at a spot he had fixed on for his headquarters.

The Allied Army marched from Paderborn on 21 June, and proceeded to the west *via* Gesecke, Soest, and Wippringshausen. Granby led with the Grenadiers, Highlanders, the 5th, 12th, 24th, and 37th Foot, and Mausberg's Regiment of Infantry. His cavalry under Colonel Harvey consisted of the Scots Greys, the 7th Queen's Own Dragoons, and the 11th Dragoons.

Two squadrons of each of these cavalry regiments, it may be noted, formed his rearguard. They took part in the march and manoeuvres which culminated in turning the flank of Soubise and menacing his rear, as has been related.

It is stated that the difficulties of Ferdinand's thirty-hours' march to effect this were greatly increased by the severity of the weather, for a tropical storm of rain burst on the troops and continued for some hours, rendering the roads almost impassable for artillery and baggage wagons. It was, in fact, morning before Granby's column was able to get clear of the camp.

Granby's force was now the left column of the five into which the Allied Army was divided. Soubise, who had chosen to consider that the movement was neither more nor less than a retreat, looked on and did nothing, yet Granby's own column actually passed within two or three miles of his camp, a fact which makes his inaction quite inexplicable. Looked at dispassionately as a manoeuvre, this march was amazing: as bold a stroke, in fact, as any general had ever attempted. Thus matters stood when the Allies suddenly made their appearance on the Plain of Dortmund at 6 p.m. on the evening of 2 July. Astonishment at first possessed the enemy, but this soon gave way to activity.

Soubise attacked one column, which was commanded by General Wütgenau. To assist in repelling this four battalions of Granby's column were detached, and the enemy were repulsed. The Allies then went into bivouac for the night, and at dawn on the morrow formed up on the Plain of Dortmund. As a matter of fact they were too weary for serious fighting, and Ferdinand did not attack that day. Meanwhile Soubise hastily sent off his baggage to Werle. On 4 July the Allies advanced to the attack, this time Granby's corps leading on the right. Soubise did not await the onslaught, but retreated. That evening the Allies occupied the former camping ground of the enemy, and had the satisfaction of commanding the rear of their retiring forces. For some time the retreat continued, till Werle was reached, in the castle of which a small post of the Allies had been left, and this post was behind the right flank of the enemy.

Here Soubise halted, pitched his camp on the very spot vacated by the Allies a few days previously, and proceeded to make his front as secure as possible. This he did by throwing up redoubts along the entire length of a ravine which extended along his front. It was a very strong position, and was one which had as a matter of fact been occupied years before by Marshal Turenne.

A few outpost affairs took place in which neither side gained much. A considerable cannonade followed, and then the Allies withdrew to their camp, Granby's corps being stationed on the extreme left at Hemmeren. The Hereditary Prince occupied the extreme right and at but a cannon-shot distance from the enemy. In his strong position Soubise might well have remained secure, but for some unexplained reason did not do so. Instead he retired by Rhüne to Soest, the Allies following him to Hilbeck.

On 12 July the Allied Army was thus posted. It was extended towards the river Lippe, Lord Granby's column being encamped at Kirchdenkern, his left in the direction of Vellinghausen, a place situated near the road from Lippstadt and Wesel, which there runs practically along the river Lippe and between that river and the Ase, which it crosses at its junction with the Lippe at Hamm.

The main body of the Allies were posted in a line, with the Salzbach in their front and with Hilbeck, Wambeln and Hohenover in their rear, counting from right to left.

The position held by Granby was evidently the most important in the disposition of the Allied Army, and it was therefore the point which De Broglie most desired to attack and, if possible, capture.

During 14 and 15 July Lord Granby was occupied in strengthening his front by means of entrenchments. He also placed obstacles across the road to Hamm to block it as much as possible, felling trees for that purpose.

On the 15th De Broglie occupied himself throughout the morning in reconnoitring the position. It is also stated that the outposts more than once engaged in desultory skirmishes. De Broglie then withdrew to his camp and for a few hours both armies remained quiet. About four o'clock in the afternoon, however, a sudden and most strenuous attempt was made on Granby by De Broglie's force alone, for it appears that that commander had given no intimation of his intention to his colleague Soubise. Advancing in three columns along the Hamm road, De Broglie's advance guard speedily drove in the German light troops ; meanwhile his centre attacked Vellinghausen, situated in the centre of Granby's position. In all haste Granby hurried to repel the attack, and by moving his left wing in an oblique direction towards the river protected the Hamm road.

The enemy at once opened fire with artillery on the camp, which was quite within range. Here the attack was opposed by the light troops, who had rallied, and who with the support of two battalions of Highlanders successfully withstood and then repulsed the assault, and, what is more to the purpose, captured a hundred officers and men of the enemy. On Granby's left, however, the position was serious, as De Broglie had succeeded in turning it, and in this direction strong supports were hurriedly sent forward by Prince Ferdinand, who had personally ridden across to Granby in order to obtain exact information as to the situation.

The Prince considered that the disposition of Granby's force was as skilful as could be hoped for under the circumstances, and ordering Vellinghausen to be held at all costs, he departed to despatch the supports or reinforcements already mentioned thither.

Accordingly, General Wütgenau was ordered to the extreme left, with an addition to his strength of three squadrons of Carabiniers; and these were posted on the Hamm road. The Prince of Anhalt, with General Elliott's cavalry, was directed on his right, and La Lippe Bückeburg, with artillery, was stationed on the left front of the centre of the army.

The combat was now most strenuous, and Granby's force, with which we have most to concern ourselves, as the 7th Queen's Own was included in it, was heavily engaged and fought, as even a French authority is compelled to confess, 'with indescribable bravery.' Some idea of what they effected may be gained by a knowledge of the fact that even before the arrival of General Wütgenau on the left the French outposts had been repulsed, and it only required the reinforcements of Wütgenau and Anhalt to complete their discomfiture; this, too, despite the fact that five strong regiments of the enemy had been sent forward to restore the battle.

The combat lasted for two hours, though spasmodic artillery fire continued till nearly ten o'clock at night.

The opposing forces passed the night thus, the French in bivouac behind a small woody hill near Granby's camp, which hill they occupied with numerous piquets. The Allies were busily engaged in making such changes in the disposition of the force as would render the centre secure should the battle be resumed, as was anticipated, on the morrow. For this purpose General Conway's command, all composed of the Guards, was brought from the extreme right of the army to the centre, while the left (Granby's) also received a very considerable accession of strength, both in cavalry, artillery, and infantry. Another factor was also brought into play to further secure the position. The column of General Spörcken, which had hitherto remained upon the other side of the river, was ordered to march till a crossing place was discovered, and then to take up a position in support of Wütgenau on the extreme left. Throughout the night, during which the outposts of the two armies were quite close to one another, continual collisions of small bodies took place.

At dawn De Broglie advanced against Wütgenau, and again without the active co-operation of Soubise.

From the woody height already mentioned a hot and galling cannonade was maintained on the extreme left of Granby's force. Granby replied with a heavy artillery fire, which did considerable execution on the enemy there, and then ordered his infantry to storm the hill. The fight for this position was short but severe, and in the upshot the height was carried. General Spörcken's command advanced and joined the troops of Wütgenau; the French attack died away; and a retreat began. Three French regiments formed the

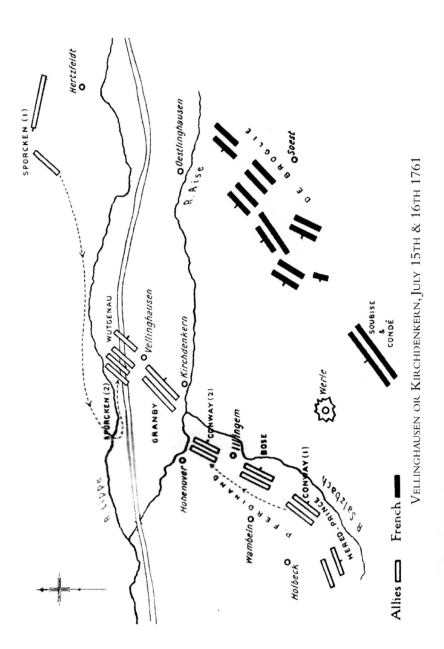

VELLINGHAUSEN OR KIRCHDENKERN, JULY 15TH & 16TH 1761

Allies ▭ French ▬

rearguard and covered the retreat; one, the Regiment de Rouge, was captured by the grenadiers under Lieut.-Colonel Maxwell; Colonel the Comte de Rouge himself, the colours, and the guns of his regiment all being the spoil of the victors. A pursuit was continued for about two miles. Meanwhile all that Soubise did was to attack in a half-hearted way the troops of the Hereditary Prince. To strengthen the forces in this part of the field, though reinforcements were hardly needed, Ferdinand detached some artillery and Hanoverians, and despatched them thither. By this time Soubise, however, had learnt of the ill-success of De Broglie's attack and his subsequent repulse, and at once himself gave way. After the battle De Broglie retreated to Oestlinghausen, and Soubise with Conde, who commanded his reserve, withdrew to Soest.

The loss of the enemy was variously estimated at from six to eight thousand killed, wounded and prisoners.

Sixty-three officers were captured, nine guns (one account says nineteen), and eight pairs of colours. The loss of the Allies did not exceed 1600 killed, wounded and missing. After the battle General Spörcken again crossed the river and proceeded to Herzfeldt, his former post. De Broglie from Oestlinghausen retreated yet further to Ervite. Soubise remained at Soest and employed the Prince of Condé to maintain the communications between his army and that of De Broglie.

The Manuscript Regimental Record omits all mention of this battle. Cannon in his History names it as Kirchdenkern. Elsewhere it is called Vellinghausen. Considering that Vellinghausen was the chief point attacked by the enemy, it is more reasonable that the battle should go by that name rather than that of the less important place.

One thing is clear, that the credit of the victory was certainly due to Granby, and it is satisfactory to find that this was freely admitted by Ferdinand, who, laying aside all personal feeling, allocated the praise to his subordinate in the most generous way.

The site of the battle of Vellinghausen was very unfavourable to the employment of cavalry, as the ground was very much broken up into fields, and the hedges formed obstacles which, while enabling infantry to be engaged, precluded anything like cavalry charges. Consequently perhaps it is that we hear practically nothing of the cavalry engaged. In the immediate pursuit, which was not prolonged more than a couple of miles, we find that only infantry were employed, and hence it is that De Broglie was enabled to draw off his beaten troops with far less inconvenience and loss than otherwise would have been the case.

Little of interest now occurred for some short time. De Broglie continued to retreat by the way of Gesecke, Buren, Haren, and Paderborn, and was closely followed up by the Allies; the vanguard, commanded by Lord Granby, being particularly alive in this direction. The retreating rearguard of the enemy was continually subjected to most harassing attacks. The main army was joined by Spörcken, and the Hereditary Prince with General Lückner again crossed the river Weser at Hameln. The enemy now re-assumed the offensive and made an attempt on the post held by the Allies at Horn. This was, however, unsuccessful, as on the approach of Granby the enemy retired.

Four days later (18 August), piquets of Beckwith's Brigade having too rashly pressed forward to attack one of the infantry posts in the rear of the enemy, got into difficulties from which they were extricated by the dashing charges of Elliot's Light Dragoons.

De Broglie, however, still continuing his retrograde movement, made a brief delay at Hoxter, which place he hastily fortified, and then crossed the river Weser, losing much of his baggage in the operation. He, however, garrisoned Hoxter, and left moreover two strong bodies of men to hold the high ground near that place. Ferdinand and Granby started to attack these, the former on the right, the latter on the left. The enemy made for the Weser and did not await the onslaughts, and Granby could do no more than open fire on them on the left, as they withdrew, while General Wütgenau paid them a similar attention on the right.

Hoxter itself next claimed the attention of the Allies, who were preparing to bombard it, Lord Granby's force being detailed for this purpose. Again the enemy did not show fight, for the garrison in some haste similarly withdrew across the Weser. The retreating Soubise had been followed up by the Hereditary Prince, with whom was General Kilmansegge. Soubise paused at Münster on 21 August, in a vain attempt to lay siege to it. The siege did not, however, take a definite form, but was converted into a blockade, and this blockade ceased on 4 September.

An attempt to bombard Hamm, for which service de Stainville was detached by Soubise, also failed. Finally Soubise was driven back towards the Rhine and de Stainville towards Cassel. The Allied Army now rested on the heights of Hoxter, that is to say on the west side of the river Weser. De Broglie lay on the east side of that river fronting Corvey.

The only mention made of the 7th Queen's Own during this period of toilsome marching and constant skirmishing is that among other troops they had behaved very well in an unnamed affair on the river Dymel in the presence of Ferdinand, who, a spectator of the combat, had been pleased to express his approbation and to bestow his thanks on all concerned. Possibly this is the affair which is thus mentioned by General Lloyd (page 230, vol. 4.):

On the 24th of August, his Serene Highness, at the head of Lord Granby's corps, and all the British troops, except the guards, proceeded by forced marches towards the Dymel, forced all the enemy's posts in that quarter, particularly at Dringlebourg, where he made upwards of 300 men prisoners, and then crossed that river; and on the 26th encamped at Hoff-Giesmar, within six leagues of Cassel, pushing forward an advanced party to Winter-Kasten. During this excursion, he left General Spörcken behind at Hoxter with the remainder of the army, to secure that part of the Weser from Hameln to the Dymel.

Meanwhile, despite the fact that he was being actively followed up by the Allies, De Broglie always managed to avoid a pitched battle. He now turned off towards the east, and detached expeditionary forces in the direction of Hildesheim, Brunswick, Wolfenbüttel, and Hanover. Contributions were everywhere levied, and all the country traversed was pillaged.

How to protect Hanover and Brunswick from siege was most important, and to avert such disasters Ferdinand changed his plans. Instead of continuing to follow up De Broglie, as the latter apparently fully-expected or hoped that he would, Ferdinand at once despatched a reinforcement to the garrison of Hanover and returned in the direction of the river Dymel and Cassel. By the Hereditary Prince a force was left under a General Oheim to watch Soubise, while he himself proceeded towards Warburg, a place to which Ferdinand and Granby were then marching. Hoxter was meanwhile still held by Spörcken.

Now during the absence of the Allies from the Dymel, and while they were on the Weser, Stainville had occupied his forces in taking up a number of posts. From these he was speedily ejected. The Allies crossed the Dymel; Stainville retired before them, till Ferdinand and Granby reached Immenhausen, where on the heights near they took up a position.

This had the effect of drawing a goodly number of De Broglie's forces from the eastern side of the Weser and in the direction of Cassel, where they collected in considerable strength.

Ferdinand then returned to his old position north of the river Dymel. The movements of the contending armies are now of a very complicated nature and need not be entered into in detail. The only event of importance was the capture, not for the first time, by the enemy of the Castle of the Sabbabourg. The strategic movements it was hoped would culminate in the defeat of Stainville, who, having advanced to Giesmar and Grebenstein (Gravenstein or Grobenstein) had again retired, this time to the heights of Immenhausen, which he in his turn occupied.

Here Ferdinand determined to attack him, and indeed could he succeed in so doing his chances of fighting a successful battle were by no means small, for the position of the enemy was certainly not a strong one. The only line of retreat towards Cassel lay through a narrow defile known as the Avenue of Wilhelmsthal, and here a retreating army could, if fairly caught, be smitten hip and thigh.

The Allies marched on the night of 17 September, crossing the Dymel in the early hours of the morning. Their design was to threaten each flank of Stainville's position with two advanced columns, these columns being commanded by the Hereditary Prince and Lord Granby respectively. The other six columns under Ferdinand (for there were eight in all) were to make a frontal attack. Again ill-fortune marred the accomplishment of the design. Stainville obtained information of the movement by some means at about 7 a.m., and this allowed him time to withdraw in safety through the defile, a manoeuvre carried out by his troops in three columns; and he reached the outskirts of Cassel in safety where he encamped at Kratzenberg. The Allies, however, not only annoyed his rear with artillery, but also captured some prisoners.

Ferdinand halted, making his headquarters at Wilhelmsthal,

Granby's force similarly occupying Weimar. The Hereditary Prince next engaged in a cavalry raid into Hesse, and succeeded in capturing 150,000 rations of oats at Fritzlar. The war now spread in the direction of Brunswick, which place was actually besieged, and this event came about thus.

Soubise and Condé had advanced towards Münster, Monsieur de Conflans had been detached against Osnabrück and Emden, and the latter place being but slenderly garrisoned was at once evacuated. As a counter-stroke the Hereditary Prince and General Hardenburg were despatched towards Monster. De Broglie crossed to the east side of the Weser when Prince Xavier of Saxony took Meppen and Wolfenbüttel and then laid siege to Brunswick.

The wealthy inhabitants of that place, as well as the reigning Duke and the Landgrave of Hesse, who was there at the time, escaped to Hamburg

Ferdinand and Granby now left their position, marching by Warburg and Hoxter to the Weser, where Granby crossed the river at Ohr by means of a pontoon bridge, and encamped at Gros-Hilligesfeldt. Wütgenau and Anhalt followed, and Ferdinand was preparing to do so when the relief of Brunswick by General Lückner was ascertained. Brunswick relieved, caused Xavier to retreat; nor was this all, for the enemy at once evacuated Wolfenbüttel and retired in haste from that district.

As for the force under Soubise, it had effected nothing more than the destruction of some depôts of stores and the capture of Emden, a by no means notable military achievement. Soubise on the relief of Brunswick retired to the Wesel. The Hereditary Prince hastened to the Weser, leaving a watching force under General Oheim to protect Hamm from any sudden attack and also to keep an eye on the movements, if any, of Soubise. He then, having obtained reinforcements from the camp at Ohr, marched to Hildesheim, where he took over the command of the troops of both General Wütgenau and the Prince of Anhalt. The prospects of the campaign had now brightened considerably, for the enemy had all been forced over to the east of the Weser, except a detachment of infantry under De Rochambeau, which still lay near Cassel. We have now arrived at the month of November, and the question arose, would there be a prolonged winter campaign, or would the opposing armies settle as usual into winter quarters? Ferdinand now made his final effort for the year.

He advanced on Eimbeck, where he found De Broglie (as regards his centre at least) too strong for attack. Ferdinand then retired towards Alfeld, despatching Granby to Vorwohle and the Hereditary Prince to Immenhausen. Marshal de Broglie, who came to the conclusion that this movement was a retreat, gave chase to the Hereditary Prince, sending the Count de Broglie in pursuit of Granby.

On the night of the 7th, just as the wearied troops of the latter had succeeded in pitching their camp at the conclusion of the toilsome march of nearly eighteen hours, Count de Broglie came up with them and promptly attacked. The outposts of Granby were driven in, but the remainder of the force turned out and with great gallantry repulsed the enemy, and moreover chased them back to their camp.

Ferdinand now assumed the offensive and made an attempt on the left flank of De Broglie. Joining Conway's division to that of Lord Granby at Vorwohle, the force moved out of camp on 9 November to Waugelstadt, while the Hereditary Prince replaced them at Vorwohle.

The events of 7 November, as far as Granby is concerned, were repeated in an almost exactly similar manner on the evening of the 9th, and again the efforts of the enemy were completely frustrated; but more than this, the effect of the repulse was to completely turn the left of De Broglie and to compel him on the night of 10 November to abandon his position, to retreat in the darkness towards Cassel, and finally to close the campaign. Nothing loath, the Allied Army prepared to follow his example.

Eimbeck was first occupied, the quarters of Granby's troops being at Salzderhelden on its south side.

Eventually Prince Ferdinand settled down for the winter, making his headquarters at Hildesheim; those of the Hereditary Prince were fixed at Minister, while Lord Granby occupied Osnabrück. The regiments in general were spread about in these places, and also in Paderborn, Hameln, Eimbeck, and Lippstadt, practically along the line of the Lippe and Dymel and to the east through the Forest of Sollingar.

Both Lord Granby and General Mostyn, the Colonel of the 7th Queen's Own, at once returned to England for the winter.

The enemy was thus disposed : Broglie went into winter quarters from Göttingen to Nordheim, also thus along the Leine.

Cannon's History does not supply many details as to the part taken by the Regiment in the complicated manoeuvres of this campaign. He tells us that after Kirch-Denkern (Vellinghausen) it 'was subsequently employed in operations which brought on slight skirmishes; but no general engagement occurred. In August it was employed on the Dymel.

In the early part of November it was engaged in dislodging a French corps from a strong camp near Escherhausen, in the Duchy of Brunswick ; and afterwards marched to Eimbeck, where another skirmish occurred. These movements were designed to surprise the French Army in dispersed quarters ; but the enemy having gained time to assemble his forces, this regiment, with several other corps, marched during the night of 7 November through a heavy snow to Foorwohle, where another skirmish occurred, and the British Dragoons evinced signal gallantry. The 7th were encamped in the snow until the following morning, when the British had another skirmish with their opponents; and they subsequently marched to the heights between Mackensen and Lithorst. When the army went into winter quarters the regiment was cantoned in East Friesland.'

The Manuscript Regimental Record makes no mention whatever of the year 1761. As to the losses sustained by the Regiment either in action or by sickness there are no numbers or lists available. Probably, however, the strength of the Regiment was very much decreased; it is an undoubted fact that, taking the Allied Army as a whole, more than one-quarter perished during the year. It is not that the actual losses in action were heavy, for they were admittedly light, but it is to hard work, hard weather, and disease that the terrible death-roll must be attributed. Another factor, though of course in a lesser degree, which must be considered was the prevalence of desertion from the ranks; this, however, was mainly to be found in the armies of our Allies, and not among the British troops.

CHAPTER 12
Battle of Wilhelmsthal
1762

We have now reached the final campaign of this war. The Allied Army remained in its winter quarters unmolested until March, when on the 10th and nth a body of 3000 French sallied forth from Göttingen and fell upon the line of the Allies at daybreak. The attack was brisk and sudden, and was not confined to one spot. The chief combat took place at Gittel, where a number fell on both sides. Kabelfeld was also attacked, but here the enemy were repulsed, losing one officer and six men prisoners.

In consequence a reinforcement of 3000 of the Allied troops was sent towards Eimbeck on the 20th, with a view to holding the garrison of Göttingen in check.

On 6 April Lückner detached 500 cavalry to Heilingstadt. The enemy becoming aware of this, issued from Göttingen with 1800 horse and 2000 foot to cut them off.

Lückner in turn becoming informed of the movement, putting himself at the head of 1600 horse, succeeded in coming up with the rear of the enemy's column while retiring on Göttingen, and attacking, killed thirty, took eighty prisoners and also captured one hundred horses. Several small affairs took place at this time, but these do not need detailed mention.

On 17 April, however, the Hereditary Prince assembled a body of troops at Unna and proceeded to Arensberg, to which place he immediately laid siege. The French Commandant offered to surrender on the 21st if he was permitted to march out with all the honours of war. and provided also that he was not relieved before that date. These terms were refused, as it was known to the Hereditary Prince that powerful forces of the enemy were marching to relieve the place.

On the night of the 18th all was quiet, but at 6 a.m. on the 19th the Allies opened both a heavy and destructive fire. At 9 a.m. the Prince offered terms which were rejected by the enemy, and firing was renewed. By noon both the castle and the town were in flames, and the place surrendered at discretion. Twenty-six cannon were captured in the castle.

This exploit caused considerable alarm in the French lines and resulted in the immediate despatch of a force across the Rhine from Cologne and Düsseldorf. This force consisted of 400 men from each battalion and 100 from each squadron. They marched on the 19th to Rattenburg, on the 20th reached Langenberg, and on the 22nd proceeded thence to Hardenberg to make room for a corps of 10,000 men assembled at Hattengen. Information, however, arrived that the Hereditary Prince had returned to his quarters, and the enemy then followed his example and retired.

The possession of Arensberg was of importance to the French for this reason: it preserved the enemy's communications between Wesel and Düsseldorf. Its capture by the Allies enabled them to raise contributions in the country of Berg.

The day after the capture of Arensberg Marshal Soubise arrived at Cassel, holding the command of the French Army on the Upper Rhine and Main in conjunction with Marshal d'Estrées. Four days later the Prince of Condé arrived at Düsseldorf to assume command of the French troops on the Lower Rhine.

On 9 May the Hereditary Prince started suddenly out to collect the contributions he had previously levied on the country of Berg. Marching from Unna with a large body of troops, he appeared at Elberfeldt before the enemy had obtained the slightest notice of his approach. Now in that neighbourhood the corps of M. de Conflans was posted, besides some other smaller bodies. The enemy retired with much precipitation, and not without loss. The Hereditary Prince then proceeded to Solingen, and having obtained hostages for the payment of the contributions, retired without any casualties. On 18 May Prince Ferdinand, escorted by a battalion of Hanoverian Guards, proceeded from Hildesheim to Pyrmont, from whence the troops which had been quartered there had departed on the previous day.

General Lückner was then posted on the Huve to await the arrival of the pontoons which were then on their way from Hanover to the Weser The Hereditary Prince next marched out and fixed his headquarters at Bulderen.

Meanwhile the French on the Upper Rhine had marked out camps and were in motion to assemble between the Fulde and the Werra. The army of the Lower Rhine was encamped at Düsseldorf, Wesel and Rees in three corps, under the command of Lieutenant-Generals Monteynard, Levi, and Saint-Chamaut.

On 24 May Lückner and Prince Frederick of Brunswick, whilst reconnoitring towards Göttingen, had a most successful little encounter with the enemy in which they captured eighty prisoners and 100 horses M. de Lahr, the French commandant, also received a mortal wound.

During these marches and skirmishes the British troops had remains: in their quarters; the time was, however, at hand for them to resume active operations in the field. Early in May transports with provisions, munitions of war, and reinforcements were despatched to Bremen from England.

Lord Granby, who it will be remembered had returned to England at the conclusion of the last campaign, now prepared to proceed to the seat of war. He was delayed by contrary winds for some days Harwich, but sailed on or about 28 May, and arrived at the Hague on 1 June. The Marquis of Granby left the Hague on 4 May for Minister, where the Hereditary Prince then was; and afterwards proceeded by Brakell and Hoxter to the headquarters of Prince Ferdinand at the Palace of Corvey on the Weser.

On 18 June the Prince reviewed the army at Brakell, whence it marched for its old position on the river Dymel, and marched with Lord Granby's Corps as usual in the vanguard.

The first object which Prince Ferdinand had in view was to cut off the communications between the garrison of Cassel and the main French Army in that district from Frankfort. This when accomplished would render the fall of Cassel, Göttingen, and Ziegenhagen merely a matter time ; and when these strongholds fell the whole of Hesse would be clear of the enemy. Granby was therefore ordered to cross to the south bank of the Dymel and to push on to Wolfshanger. Here he dislodged some of the enemy and occupied the Heights of Volksmissen, the enemy on this occasion forming a part of De Stainville's troops. Almost simultaneously an attack was made on Stainville's right by a force commanded by Lord Frederick Cavendish, which had been despatched thither with that object in view. Cavendish came into collision with the enemy near Gravenstein (Grebenstein or Grobenstein), and after nearly succeeding in capturing De Stainville himself, as well as the Duke de Cogni and De Castries, he proceeded to retake the Sabbabourg. Soubise and d'Estrées now advanced in force against Granby, who in consequence retired across the Dymel, having achieved the object he had in view, which was to disturb the enemy to the northwest of Cassel. A party was now sent to occupy the passes leading from the south of the Dymel into Hesse, and another to secure the communications with General Lückner.

At this time and until the night of the 24th of June the 7th Queen's Own had been in camp, first at Brakell and afterwards on the heights of Tissel. They were brigaded with the 11th Dragoons, and the brigade was commanded by Lieut.-Colonel George Lawson Hall of the 7th.

The only information to be obtained as to the strength of the Regiment at this period comes from a paper dated 1762, 2 June, to be found among the Stowe M. S. S. in the British Museum, and as the strength there given is probably that of the Regiment at the beginning of the campaign, at any rate on paper, of that year, it may well be inserted here:

7th Mostyn's. In Germany, strength 28 commissioned and warrant officers; 24 non-commissioned officers; 243 men fit for duty; 16 sick; 311 total officers and men; 47 wanting to complete; 24 contingent men. Total 382.

The same paper tells us that the Light Troop had 6 officers commissioned and warrant; 6 non-commissioned officers; 74 men fit for duty; 2 sick; total 88 officers and men; 41 wanting to complete, 5 contingent men; 134 total. It will be remembered that the Light Troop had been left behind in England.

A fact in connection with this campaign, which the reader will hardly fail to have noticed, is the absence of Marshal De Broglie.

The explanation is this: thanks to court intrigues mainly, and also to the mortification of the French, in consequence of their serious defeat at Vellinghausen, De Broglie, who, by the way, was admittedly the greatest soldier then possessed by that nation, had fallen under the severe displeasure of both the King and those surrounding him. The Marshal was, in fact, sent into exile, and Soubise would have shared the same fate had it not been for the powerful influence of Madame de Pompadour, with whom he was a great favourite. Thus the strong man and skilled commander was disgraced, while Soubise, his inferior in war, was continued in his command with Marshal d'Estrees appointed as his colleague.

The strategy of Prince Ferdinand had at length succeeded in drawing the enemy into a position which promised him an opportunity of fighting favourable battle. The French were now behind Gravenstein (otherwise Grebenstein, Gröbenstein, and Groebenstein). Ferdinand would have preferred perhaps that they should have occupied the heights of Immenhausen, but this he was unable to accomplish. What he proposed to do was to fight, and having entangled the enemy in the pass or gorge at Wilhelmsthal, to destroy them. He accordingly gave orders for the Allies to cross the Dymel early on the morning of 24 June. He had sent General Lückner across the Weser on the preceding night, leaving a part of his force behind near the river Leine in order to conceal his movement from Prince Xavier of Saxony. Having crossed, Lückner advanced in the rear of the French right, which was commanded by De Castries. The garrison of the Sabbabourg was strengthened.

Granby, who on this occasion commanded what was technically the reserve, crossed the Dymel on the morning of 24 June ahead of Prince Ferdinand, and posted his men on the Dürrenberg. Here, though he could have been hindered, if not prevented, from taking up a position, he was unmolested by the enemy; and awaiting developments Granby's force remained absolutely in view of the tents of the French camp. Ferdinand's main body marched at 4 a.m. on 24 June, with General Spörcken's command on its extreme left; it being intended that Spörcken and Lückner should act in concert. Ferdinand was, however, too late in arriving, and moreover did not proceed far enough before he formed in fighting line. Lückner appears to have timed his movement with great exactitude up to a certain point, and having arrived within striking distance of De Castries near Holzhausen and Mariendorff, attacked the enemy and drove them in on the French right. Granby set his troops in motion and also arrived to time at Fürstenwald. Here the guns of Lückner were audible, and believing, incorrectly, that they were those of Ferdinand and that the French centre was being assailed, Granby despatched a small body of men with two guns into the wood of Fürstenwald, and then with the remainder of his troops rapidly marched on to Wilhelmsthal, hoping to prevent the escape of the enemy.

WILHELMSTHAL, 24TH JUNE 1762

This march was made quite close to the left flank of the French. Meanwhile the infantry of the enemy were already retreating, and retreating in haste, for finding Lückner on their right flank and Granby appearing at Fürstenwald on their left rear, they had taken alarm and were now hastening towards Wilhelmsthal. Of course, had Prince Ferdinand been within a distance which would have permitted him to press an attack on their centre, a great disaster must inevitably have befallen them, but he was too far off to strike and strike home.

Granby by his rapid march by the left cut off a detached rearguard which was occupied in covering the retreat of the enemy into and through the pass. These Granby was preparing to charge, when they in their turn by a, movement round the wood of Fürstenwald succeeded in gaining the of his reserve, and captured two guns and a grenadier company of an infantry regiment.

Granby was now thus placed: in his rear was the rearguard of the enemy, in his front the whole of its left flank, though this left flank was in retreat, while nothing was either to be seen or heard of the Allied centre. Matters looked most unfavourable for Granby, that is to say if the enemy's left flank should rally. Granby, however, swiftly made up his mind as to what course of action he should pursue. Neglecting the enemy's left flank, he turned back on the rearguard, and after a strenuous conflict overthrew it. De Stainville was completely defeated, and was driven back on the right of Ferdinand's troops, which had by this time arrived or were arriving. The result was the surrender of three entire regiments to the 5th Foot, besides others which had already been captured A general flight of the enemy took place. De Stainville, with a very small portion of his command, escaped; Soubise and d'Estrées were chased through the pass at Wilhelmsthal and right up to Cassel itself, with a loss of 6000 men, of whom 2700 were prisoners; 170 officers were captured, and also one standard, six colours, two guns and the whole of the baggage, &c. On the left General Spörcken inflicted some loss on the enemy and took two guns. The Allied Army suffered but few casualties.

With regard to the part taken by the 7th Queen's Own Dragoons in this battle Cannon writes as follows:

They left their camp before daybreak on the morning of the 24th of June, 1762, and, having crossed the Dymel, advanced against the French camp at Groebenstein. The enemy was surprised, and made a precipitate retreat upon Cassel, with the loss of their tents and baggage; and one division being surrounded in the woods of Wilhelmsthal, surrendered. The Seventh pursued the French in the direction of Cassel and took several prisoners; they subsequently encamped near Holtzhausen.

The Manuscript Regimental Record is even more brief:

1762. The Regiment was present at the Battle near Gravenstein, 24 June. The Regiment does not appear to have sustained any loss in this Action, see return inserted in the *London Gazette*, 10 July, 1762.

The losses of the Allies are estimated at 700 men, of which 450 belonged to the troops under the command of Granby. Officially, the name by which this battle is known is Wilhelmsthal.

When the failure in part of both Spörcken and Lückner is considered, and the late arrival of Prince Ferdinand is remembered, the credit of any success in this somewhat disappointing battle assuredly lies with Lord Granby. The remainder of the campaign, as far as these pages are concerned, may be passed over.

That the Regiment was for the next few months actively employed, chiefly on detached services which entailed continuous marches in every direction, we are told, but no details are given. There are no records that we have been able to obtain, despite careful search, to furnish even the briefest list of the places visited, the objects to be attained, or whether the operations were or were not successful in their issue.

The Preliminaries of Peace were signed on 3 November, though news of the signature did not reach Prince Ferdinand for some days. On 16 November the enemy began to march homewards, and three days later the Allies went into winter quarters in the bishoprics of Munster and Osnabrück, and along the frontiers of Hanover.

One event towards the end of the campaign alone need be mentioned, and that was the capture of Cassel, a stronghold that had been in the possession of the enemy since 1760.

According to a return furnished towards the end of the war the strength of the 7th Queen's Own Dragoons was 516 officers and men. It is, however, open to doubt whether this is correct, and in the face of there being no record of any drafts sent to reinforce the strength of the Regiment during the year, we are more inclined to accept the strength on active service given earlier in this chapter, as the numbers here given were obtained by adding together the totals of the troops with Mostyn in Germany and of the Light Troop which remained at home. In all probability the efficient strength of the Regiment on this campaign was little more than 300 officers and men.

CHAPTER 13

Home Service
1764-1792

When peace was declared the 7th Queen's Own Dragoons recorders, in February 1763, to return to England. Accordingly they marched through Germany and Holland to their port of embarkation. Williamstadt, where vessels awaited them. The date of embarkation is not given, but the Regimental Record tells us that on 11 March:

> The 7th Dragoons on arrival from Germany at Harwich or Purfleet to disembark and march as follows:
> 3 Troops to Chelmsford & Springfield
> 3 Troops to Colchester

Reductions in establishment were now the order of the day.

The Light Troop, which had been stationed at Canterbury since 7 October 1762, and had been employed in patrol duty from Dove: to Hastings, was on 14 January 1763 ordered to be reduced to 1 captain. 2 lieutenants, 2 cornets, 3 sergeants, 3 corporals, 2 drummers, and 60 privates. This reduction was but a preliminary to its complete disbandment, for which an order arrived on 25 March, and the disbandment. accordingly took place forthwith.

On the same date the remaining six troops of the Regiment were reduced to the strength per troop given: 1 captain, 1 lieutenant, 1 cornet, 1 quartermaster, 2 sergeants, 1 corporal, and 28 private men. Eight men in each troop were, however equipped as light dragoons, while the remaining twenty remained as heavy dragoons.

On 23 April the troops at Chelmsford and Springfield were ordered to march on the 29th to Witham, Kelvedon, and Coggeshall and Braintree, while the Colchester troops on 11 May were ordered to proceed on or before the 23rd to Maldon, Halstead, and Thaxted and Dunmow respectively. Next day (24 May) the whole Regiment was ordered to return to its former quarters, the three former Colchester troops proceeding, however, to Chelmsford and Springfield, while the others went to the quarters that had previously been occupied by those at Colchester: it was in fact an interchange.

Here the three troops remained until the Chelmsford election in December, when they were ordered, as was then customary, to leave that place and to proceed during the election time to Colchester, and after the declaration of the poll they were to return.

When in due course the dates of the county election held at Colchester were fixed, the three troops there were for that time shifted to Sudbury, on the borders of Suffolk.

On 3 January, 1764, a detachment consisting of 1 captain, 2 subalterns, and 60 men, with a due proportion of non-commissioned officers, was sent to Ipswich on revenue duty.

On 9 April the whole Regiment was concentrated and proceeded to Hyde Park to be reviewed by King George III. His Majesty was pleased on the occasion to compliment the 7th Queen's Own on their appearance and discipline. On the 14th the Regiment, which had returned to Chelmsford and the district, received orders to march on the 24th as follows: three troops to Worcester and three troops to Hereford. But by a supplementary order it was notified that they were to go instead to Worcester and Pershore till 30 May. Here the Regiment remained till 7 March, when the Worcester troops were despatched to Hereford in consequence of the assizes being held in the former city, after which the Regiment was to concentrate at Worcester.

On 21 March the Regiment was ordered to march in two divisions to the south coast, two troops to Lewes and one to Brighton, two troops to Chichester and one to Havant. The march was to begin on 3 April and to be completed by 17 April. On 5 April, however, these orders were varied as follows: having arrived at Reading, two troops of the 2nd Division were ordered to Devizes, there to remain till relieved by the 1st Dragoons. The other four troops were to proceed as follows: one to Colchester, two to

Lewes and one to Brighton. On 20 April the Devizes troops were despatched to Chichester and Havant. Six days later the quarter; were again partially changed.

Two troops remained at Chichester, two at Lewes, one at Eastbourne and Pevensey, and one at Hastings. The Regiment was now allowed to remain quiet for a few months. On 19 October all the troops were concentrated at Lewes for review, and were afterwards ordered to proceed, one troop to Hastings, one to Brighton, two to Chichester, and two to remain at Lewes.

Meanwhile the detachment on revenue duty was still absent. It was, however, recalled on 14 March 1766, when the whole Regiment was ordered to Lewes and Brighton, to assemble at those places on the 31st.

On 11 April the Regiment was ordered to Lambeth, the marches to take place between the 22nd and 26th of the month. At Lambeth the Regiment remained until 12 May, when it was again dispersed, two troops being ordered to Northampton, two to Leicester, and one each to Wellingborough and Loughborough. On 18 December the Loughborough troop was withdrawn and posted at Northampton. No further change occurred in the disposition of the Regiment until 16 March 1767, when the 7th Queen's Own were ordered to assemble at Leicester by Thursday the 26th.

Cannon tells us that during the preceding year while at Leicester and Northampton the drummers on the establishment of the Regiment were ordered to be replaced by trumpeters. It would be expected that some notice of this important change would have been met with in the Manuscript Regimental Record, but no trace of it is to be discovered therein. Similarly the order in 1764 for the Regiment to be mounted on long-tailed horses and for epaulettes to be worn on the left shoulder in lieu of aiguillettes and for the boots of the men to be of a lighter description than formerly, is also omitted, though given in Cannon.

The removal of Lieut.-General Mostyn in May 1763 from the Colonelcy of the Regiment to that of the First Dragoon Guards and the appointment of Major-General Sir George Howard, K. B., to succeed him in the 7th Queen's Own, is likewise unrecorded.

Sir George Howard had previously been Colonel of the 3rd Foot (the Buffs). General Mostyn had now obtained the command of the Regiment which he had for long most ardently desired, and for which he for some years had been waiting. Judging from his letters, we are inclined to think that Mostyn almost resented his appointment to the 7th Queen's Own, and accepted the Colonelcy most unwillingly. (See Register of Officers.)

For the years 1765 and 1766 some interesting particulars have been obtained, and are here inserted. 8 October 1765.—Among other regiments the 7th were ordered as follows at reviews:

Regimental swords are to be used by the officers of Dragoons instead of fusils when the Regiments dismount, or when they parade on foot. After the word of command Dismount, before the men face to the left, the swords of the dragoons are to be fixed on the saddle, by placing the hilt on the pummel, and fixing it with the firelock strap, the point to be under the middle cloak strap. This is to be done without any word of command, and immediately after dismounting. In performing the horse evolutions, and in taking ground by squadron, or by divisions, The movements not to be made on a gallop, but on a brisk trot, or walk as occasion requires. No galloping, except at the charge, and which should not continue above 100 paces. When Regiments march on foot by a reviewing general, they are to march with fixed bayonets.

22 October 1765.—It had been found that there was great want of uniformity in the honours with which the Major Generals at reviews were received. Among other regiments the 7th Dragoons were ordered (notwithstanding any former regulations) in future that the 'Drummers are to beat two ruffles and all officers to salute (the Standards) excepted.'

December 1765.—The recruiting officer of any Dragoon regiment is forbidden to give more than 'a guinea and a crown' as bounty money. For infantry the sum was fixed at 'a guinea and a half.'

On 17 January 1766 the order for Dragoon horses to have 'full tails without the least part having been cut off' was issued.

11 March 1766.—It was ordered that in the Review Returns in dragoon regiments, 'for the farriers, neither firelock, pistols or swords are to be accounted for—and for the drummers no other arms but a sword and pistol for each.'

28 June 1766.—On this date His Majesty gave orders that 'the Regiments of Dragoon Guards and Dragoons shall have trumpeters to each troop, instead of the drummers.' Two master trumpeters, by name Mr. Abingdon and Mr. Willis were therefore appointed to undertake 'the perfecting of some the present drummers of each corps in their duty as trumpeters, and a proper place at the Horse Guards and instruments are procured for the purpose. The men who are sent by the several Regiments are to be quartered in the neighbourhood of that place and to be under the command of the non-commissioned officers who are ordered to attend. The masters are to receive 4 guineas for each man that is perfected in the regimental duty, as a trumpeter. This is to be advanced by the respective regiments when demanded and charged in the contingent account.

> At present the Queens Regiment of Dragoon Guards is to send two and they may set out immediately. By taking a copy of a beating order they will get billets on the road, and when arrived on applying at the war office they will receive further orders.
> The Regiments are to subsist them as far as London and give directions to their respective agents relative to advancing them their subsistence while continuing on this duty. You will please let me know how many of the present drummers you propose to keep as trumpeters, and how many it will be necessary to discharge (as not being proper for that purpose) with the reasons.

This order was sent to the 7th Queen's Own Dragoons among other regiments. It was proposed to recommend the discharged drummers, if of good character, &c, for the bounty of Chelsea Hospital.

9 October 1766.—A letter about the Cavalry Review in the autumn is rather interesting. It is as follows:

> *Sir*, His Majesty (to prevent the ill effects to horses from being too much heated by violent movements soon after they are taken up from grass, and that they have only necessary gentle exercise) has given directions to the reviewing generals that the several regiments of horse, dragoon guards and dragoons are not to practice any manoeuvres at the Autumn review with rapidity.
> *E. H.*, 9 Oct. 1766

A note as to the price of Commissions at this date is of interest. 31 January 1766.

Price of Commissions

Dragoon Guards and Dragoons

Rank	Price	Difference
Lieut.-Colonel	£4700	£1100
Major	£3600	£1100
Captain	£2500	£1100
Captain-Lieutenant	£1400	£250
Lieutenant	£1150	£150
Cornet	£1000	£100
		£4700

All Commissions in the Dragoons were calculated at ten years' purchase.

	Half-Pay per Diem		Half-Pay per Annum			Half-Pay at ten years' purchase		Price of Commission	Difference	
	s	d	£	s	d	£	s	£	£	s
Lieut.-Colonel	10	0	182	10	0	1825	0	4700	2875	0
Major	8	0	146	0	0	1460	0	3600	2140	0
Captain	5	6	100	7	6	1003	15	2500	1496	5
Lieutenant	3	0	54	15	0	547	10	1150	602	10
Cornet	2	6	45	12	6	456	5	1000	543	15

To return to the movements of the Regiment, which was now at Nottingham. On 7 May 1767 it was ordered to march to York, the date of the march being fixed for 11 May, and of arrival for 16 May. At York the 7th Queen's Own remained until 15 February 1768, on which date orders to proceed to Durham were received, the march to begin on 26 February.

On 25 March they were ordered to march on 6 April from Durham to Berwick and to proceed thence to North Britain. We are not told where in Scotland the Regiment was quartered, either in Cannon or in the Manuscript Record, but we know that an order was received 28 February 1769 to the effect that on arrival from North Britain they were to march three troops from Coldstream to Coventry from 22 March to 14 April, and the other three troops from the same place to the same from 24 March to 15 April. We may conclude therefore that the Regiment had concentrated and was marched south in two divisions. On 15 May three troops were sent from Coventry to Warwick, and on 23 October two of these troops were shifted to Stratford-on-Avon. No other change of quarters took place until 2 February 1770, when the same two troops were moved back from Stratford-on-Avon to Warwick.

On 17 March we find that the whole Regiment was ordered south, the three troops at Warwick marching from 7 to 13 April and those at Coventry from 7 to 14 April; their destination being Newbury and its suburb Speenham Land. The stay of the Regiment here was but brief, as on 26 April they were ordered to proceed from 5 May to Dorchester, Axminster, Bridport, Blandford, Yeovil, and Sherborne; one to each place.

There is no other entry for 1770. A slight change of quarters, however, took place on 4 January, 1771, when the troops at Axminster, Bridport, and Yeovil were ordered to join the troops at Dorchester, Sherborne, and Blandford respectively. During the month of February another slight change of quarters took place, one of the Dorchester troops being sent to Blandford and another to Sherborne. This was, it would seem, a preliminary to the assembly of the Regiment at Salisbury, which took place on 12 March, after which the six troops were quartered at Salisbury, Fisherton, Harnham (?) and Milford, though in what proportions is not stated. On 29 March the Regiment received orders to march on 2 April as follows: three troops to Henley and three troops to Maidenhead.

Within a month they were again shifted; marching so as to arrive on 30 April, two troops to Putney, one to Fulham, and three to Richmond-Here the Regiment remained till some time in June, when the entire six troops were ordered to Canterbury on coast duty, and were so engaged till April 1772—the exact dates are not stated. From May to October 1772 we find the Regiment in new quarters, and thus dispersed : one troop each at Daventry, Newport Pagnall, Wellingborough, and Market Harborough, the remaining two being at Boston. From November 1772 to March 1773 the Daventry troop was moved to Northampton.

On 10 March 1773 the Regiment, having assembled, marched to York, arriving on 22 March. Four days later they started for Berwick-on-Tweed, which they reached on 7 April. Here there is a gap until June, when we find all six troops at Haddington. From September 1773 to May 1774 the Regiment was stationed at Linlithgow, at the end of which period it again returned to England, for we read that in June all six troops were at Preston. There is absolutely no information in any detail of the movements of the Regiment at this period; we shall therefore be as brief as possible, merely recording the information obtainable from the Regimental Manuscript.

September 1774. Six troops at Manchester ; December 1774 to March 1775, three troops at Manchester, two at Warrington, and one at Wigan. April 1775 to March 1776, six troops at Worcester, Pershore, and Newcastle-in-Emlyn (Carmarthenshire); the numbers at each place, however, are not stated. From May 1776 to July 1777 the Regiment was sent to Sussex on preventive or coast duty, being distributed between Arundel, Lewes, Brighton, Chichester, and Havant. Coast duty in Norfolk and Suffolk followed from August to December 1777, three troops being stationed at Norwich and three at Ipswich. Whether the Regiment was moved thence in January and February 1778 we do not know, but it was back again in the same quarters in February 1778.

In the month of August of this year we find the Queen's Own Dragoons encamped near Bury St. Edmunds, five troops being at the Stow camp and one at Hengrave. Three other regiments of dragoons were also present and two battalions of militia, the whole being under the command of Major-General Warde. There were quite an unusual number of camps this year. A large one was formed near Cork on 25 July and 'others were forming in the most convenient stations throughout Ireland.' A second was formed at Winchester; where the troops were reviewed by the King with much pomp and circumstance on 29 September. A camp formed at Salisbury broke up on 7 October. On 10 October, after visiting Lord Petre at Thorndon Hall, Essex, the King and Queen were present at a review of the troops in camp at Warley. Lastly, on 3 November the King also reviewed the troops encamped at Coxheath near Sevenoaks, having visited Lord Amherst at Montreal prior to the review, and proceeding to Leeds Castle after its conclusion.

LIEUTENANT–COLONEL THOMAS BLAND, 1771

In March 1779 we find two troops at Newmarket, two at Hadleigh and one each at Boxford and Lavenham. During April the Colonel of the Regiment, Lieut .-General Sir George Howard, C.B., was removed to the command of the First Dragoon Guards vice Mostyn deceased. He was succeeded in the Colonelcy of the 7th Queen's Own Dragoons by Major-General Sir Henry Clinton, K. B. This promotion is not noticed in the Regimental Manuscript. In the same month a most important regimental change took place. Firstly, owing to the outbreak of the American War the strength of the Regiment was augmented. Secondly, in April the eight men per troop that had been equipped as Light Dragoons since 1763 were incorporated with detachments from the 2nd, 3rd, 15th, and 16th Dragoons into a separate regiment, which was numbered as the 21st Light Dragoons. This new regiment was the second cavalry regiment in the British service to be numbered 21. The first was raised by the Marquis of Granby in 1760 and was known as the 21st or Royal Windsor Foresters. It was disbanded in 1763.

The second was the regiment raised as already mentioned in 1779. The first Colonel was General Douglas. During its brief existence—for it was disbanded on 1 June 1783—the regiment served only in England, *viz.* for three years in the Midlands at Manchester in 1781, Colchester 1782, then at Lanham Camp at Canterbury from November 1782 till it was disbanded. The third 21. was re-raised in 1794, and after distinguishing itself on active service was disbanded at Chatham in 1819. The fourth and last 21., which still exists as the 21. (Empress of India's) Lancers, was raised in India from the Bengal European cavalry after the Mutiny. This regiment was equipped as Hussars in 1862 and as Lancers in 1897.

The 7th Queen's Own were now sent to Scotland, though the date of their departure thither is not given. We, however, find that three troops were at Montrose and three at Kilmarnock in September 1779.

In the following year all the six troops were assembled at Montrose in May and proceeded to Musselburgh in July. From September 1780 to April 1781 the Regiment was concentrated at Haddington. The 7th Queen's Own was now ordered back to England, and we find that from May 1781 to April 1782 the

Regiment was quartered at Durham. In the following June they had marched into Wiltshire and were posted at Salisbury, proceeding thence to Marlborough in December. In June 1783 the Regiment moved its quarters to Newbury, and in December proceeded to Reading; where it remained until June 1784. At that date and until December it was quartered at Hounslow, then from January to June 1785 at Croydon.

It must here be noted that an important change in equipment took place in 1784 just prior to the stay at Hounslow, for the Regiment, hitherto Heavy Dragoons, was then equipped as Light Dragoons. The details of this change will be fully dealt with in the chapter on uniforms. For the present it is only needful to state that helmets replaced the old cocked hats, the boots, saddles, belts and other equipments were ordered to be of a much lighter kind of make. A carbine replaced the old short musket, and as a weapon, owing to the shortness of its barrel, it was by no means an improvement. In addition the standard of height both for men and horses was reduced. During its stay at Hounslow the Regiment was mainly employed in escort duty for the Royal Family, having detachments for this purpose spread about in the villages along the road from London to Windsor.

Before leaving Hounslow for Croydon in 1784 the Regiment was reviewed by the King. No details of this review are to be obtained, the Gentleman's Magazine, which usually reported such functions, omitting in this case to make any mention thereof. Cannon, however, tells us that the 'appearance and discipline' of the Regiment 'procured the expression' of the King's approbation. The Manuscript Record does not chronicle the review.

Apparently the Regiment only remained about six months in Croydon, and must have occupied billets, as the cavalry barracks were then un-built in that town. They were constructed in 1794 at a cost of £16,188 5s. 9¼d., the contractor being a Mr. Thomas Tomlins. Hence probably it is that in the month of July the Regiment was sent on revenue duty to the south coast, being posted until March 1786 at Chichester, and places in the neighbourhood.

In May 1786 the entire Regiment marched to Guildford, where remained till July, when orders were received to proceed to Maidstone. The Regiment next lay at Canterbury and possibly occupied the old St. Gregory's Barracks, which were succeeded by the King's Barracks erected in 1798 at a cost of £31,175 3s., and built by William Baldock.

The Record states that the Regiment remained at Canterbury until March 1787. When it left that city is not clear, but it was at Greenwich in May, and at Norwich in June.

Here the Regiment remained for an entire year, when it moved to Nottingham and the neighbouring places, arriving in July 1788 and being quartered there till August. In October 1788 we find the 7th Queen's Own at Loughborough, and in December back again at Nottingham, where it remained till April 1789.

In June the Regiment was quartered at Staines, and from August 1789 to April 1790 at Hounslow.

Next came a turn of duty on the south coast, and from June 1790 to February 1791 they were quartered at Lewes. In April 1791 the station was Brighton, then from June 1791 to February 1792 Canterbury. Greenwich followed in April, and from June to December 1792 Nottingham again.

Foreign active service was now about to engage the Regiment, the history and events of which are treated in the second volume of this history.

LEONAUR

ALSO FROM LEONAUR
AVAILABLE IN SOFTCOVER OR HARDCOVER WITH DUST JACKET

CAPTAIN OF THE 95th (Rifles) *by Jonathan Leach*—An officer of Wellington's Sharpshooters during the Peninsular, South of France and Waterloo Campaigns of the Napoleonic Wars.

BUGLER AND OFFICER OF THE RIFLES *by William Green & Harry Smith* With the 95th (Rifles) during the Peninsular & Waterloo Campaigns of the Napoleonic Wars

BAYONETS, BUGLES AND BONNETS *by James 'Thomas' Todd*—Experiences of hard soldiering with the 71st Foot - the Highland Light Infantry - through many battles of the Napoleonic wars including the Peninsular & Waterloo Campaigns

THE ADVENTURES OF A LIGHT DRAGOON *by George Farmer & G.R. Gleig*—A cavalryman during the Peninsular & Waterloo Campaigns, in captivity & at the siege of Bhurtpore, India

THE COMPLEAT RIFLEMAN HARRIS *by Benjamin Harris as told to & transcribed by Captain Henry Curling*—The adventures of a soldier of the 95th (Rifles) during the Peninsular Campaign of the Napoleonic Wars

WITH WELLINGTON'S LIGHT CAVALRY *by William Tomkinson*—The Experiences of an officer of the 16th Light Dragoons in the Peninsular and Waterloo campaigns of the Napoleonic Wars.

SURTEES OF THE RIFLES *by William Surtees*—A Soldier of the 95th (Rifles) in the Peninsular campaign of the Napoleonic Wars.

ENSIGN BELL IN THE PENINSULAR WAR *by George Bell*—The Experiences of a young British Soldier of the 34th Regiment 'The Cumberland Gentlemen' in the Napoleonic wars.

WITH THE LIGHT DIVISION *by John H. Cooke*—The Experiences of an Officer of the 43rd Light Infantry in the Peninsula and South of France During the Napoleonic Wars

NAPOLEON'S IMPERIAL GUARD: FROM MARENGO TO WATERLOO by *J. T. Headley*—This is the story of Napoleon's Imperial Guard from the bearskin caps of the grenadiers to the flamboyance of their mounted chasseurs, their principal characters and the men who commanded them.

BATTLES & SIEGES OF THE PENINSULAR WAR by *W. H. Fitchett*—Corunna, Busaco, Albuera, Ciudad Rodrigo, Badajos, Salamanca, San Sebastian & Others

LEONAUR

ALSO FROM LEONAUR

AVAILABLE IN SOFTCOVER OR HARDCOVER WITH DUST JACKET

WELLINGTON AND THE PYRENEES CAMPAIGN VOLUME I: FROM VITORIA TO THE BIDASSOA *by F. C. Beatson*—The final phase of the campaign in the Iberian Peninsula.

WELLINGTON AND THE INVASION OF FRANCE VOLUME II: THE BIDASSOA TO THE BATTLE OF THE NIVELLE *by F. C. Beatson*—The second of Beatson's series on the fall of Revolutionary France published by Leonaur, the reader is once again taken into the centre of Wellington's strategic and tactical genius.

WELLINGTON AND THE FALL OF FRANCE VOLUME III: THE GAVES AND THE BATTLE OF ORTHEZ *by F. C. Beatson*—This final chapter of F. C. Beatson's brilliant trilogy shows the 'captain of the age' at his most inspired and makes all three books essential additions to any Peninsular War library.

NAVAL BATTLES OF THE NAPOLEONIC WARS *by W. H. Fitchett*—Cape St. Vincent, the Nile, Cadiz, Copenhagen, Trafalgar & Others

SERGEANT GUILLEMARD: THE MAN WHO SHOT NELSON? *by Robert Guillemard*—A Soldier of the Infantry of the French Army of Napoleon on Campaign Throughout Europe

WITH THE GUARDS ACROSS THE PYRENEES *by Robert Batty*—The Experiences of a British Officer of Wellington's Army During the Battles for the Fall of Napoleonic France, 1813.

A STAFF OFFICER IN THE PENINSULA *by E. W. Buckham*—An Officer of the British Staff Corps Cavalry During the Peninsula Campaign of the Napoleonic Wars

THE LEIPZIG CAMPAIGN: 1813—NAPOLEON AND THE "BATTLE OF THE NATIONS" *by F. N. Maude*—Colonel Maude's analysis of Napoleon's campaign of 1813.

BUGEAUD: A PACK WITH A BATON *by Thomas Robert Bugeaud*—The Early Campaigns of a Soldier of Napoleon's Army Who Would Become a Marshal of France.

TWO LEONAUR ORIGINALS

SERGEANT NICOL *by Daniel Nicol*—The Experiences of a Gordon Highlander During the Napoleonic Wars in Egypt, the Peninsula and France.

WATERLOO RECOLLECTIONS *by Frederick Llewellyn*—Rare First Hand Accounts, Letters, Reports and Retellings from the Campaign of 1815.

LEONAUR

ALSO FROM LEONAUR

AVAILABLE IN SOFTCOVER OR HARDCOVER WITH DUST JACKET

THE JENA CAMPAIGN: 1806 *by F. N. Maude*—The Twin Battles of Jena & Auerstadt Between Napoleon's French and the Prussian Army.

PRIVATE O'NEIL *by Charles O'Neil*—The recollections of an Irish Rogue of H. M. 28th Regt.—The Slashers— during the Peninsula & Waterloo campaigns of the Napoleonic wars.

ROYAL HIGHLANDER by *James Anton*—A soldier of H.M 42nd (Royal) Highlanders during the Peninsular, South of France & Waterloo Campaigns of the Napoleonic Wars.

CAPTAIN BLAZE *by Elzéar Blaze*—Elzéar Blaze recounts his life and experiences in Napoleon's army in a well written, articulate and companionable style.

LEJEUNE VOLUME 1 by *Louis-François Lejeune*—The Napoleonic Wars through the Experiences of an Officer on Berthier's Staff.

LEJEUNE VOLUME 2 by *Louis-François Lejeune*—The Napoleonic Wars through the Experiences of an Officer on Berthier's Staff.

FUSILIER COOPER *by John S. Cooper*—Experiences in the 7th (Royal) Fusiliers During the Peninsular Campaign of the Napoleonic Wars and the American Campaign to New Orleans.

CAPTAIN COIGNET *by Jean-Roch Coignet*—A Soldier of Napoleon's Imperial Guard from the Italian Campaign to Russia and Waterloo.

FIGHTING NAPOLEON'S EMPIRE by *Joseph Anderson*—The Campaigns of a British Infantryman in Italy, Egypt, the Peninsular & the West Indies During the Napoleonic Wars.

CHASSEUR BARRES *by Jean-Baptiste Barres*—The experiences of a French Infantryman of the Imperial Guard at Austerlitz, Jena, Eylau, Friedland, in the Peninsular, Lutzen, Bautzen, Zinnwald and Hanau during the Napoleonic Wars.

MARINES TO 95TH (RIFLES) by *Thomas Fernyhough*—The military experiences of Robert Fernyhough during the Napoleonic Wars.

HUSSAR ROCCA by *Albert Jean Michel de Rocca*—A French cavalry officer's experiences of the Napoleonic Wars and his views on the Peninsular Campaigns against the Spanish, British And Guerilla Armies.

SERGEANT BOURGOGNE by *Adrien Bourgogne*—With Napoleon's Imperial Guard in the Russian Campaign and on the Retreat from Moscow 1812 - 13.

LEONAUR

ALSO FROM LEONAUR

AVAILABLE IN SOFTCOVER OR HARDCOVER WITH DUST JACKET

A JOURNAL OF THE SECOND SIKH WAR by *Daniel A. Sandford*—The Experiences of an Ensign of the 2nd Bengal European Regiment During the Campaign in the Punjab, India, 1848-49.

LAKE'S CAMPAIGNS IN INDIA by *Hugh Pearse*—The Second Anglo Maratha War, 1803-1807. Often neglected by historians and students alike, Lake's Indian campaign was fought against a resourceful and ruthless enemy-almost always superior in numbers to his own forces.

BRITAIN IN AFGHANISTAN 1: THE FIRST AFGHAN WAR 1839-42 by *Archibald Forbes*—Following over a century of the gradual assumption of sovereignty of the Indian Sub-Continent, the British Empire, in the form of the Honourable East India Company, supported by troops of the new Queen Victoria's army, found itself inevitably at the natural boundaries that surround Afghanistan. There it set in motion a series of disastrous events-the first of which was to march into the country at all.

BRITAIN IN AFGHANISTAN 2: THE SECOND AFGHAN WAR 1878-80 by *Archibald Forbes*—This the history of the Second Afghan War-another episode of British military history typified by savagery, massacre, siege and battles.

UP AMONG THE PANDIES by *Vivian Dering Majendie*—An outstanding account of the campaign for the fall of Lucknow. *This is a vital book of war as fought by the British Army of the mid-nineteenth century, but in truth it is also an essential book of war that will enthral military historians and general readers alike.*

BLOW THE BUGLE, DRAW THE SWORD by *W. H. G. Kingston*—The Wars, Campaigns, Regiments and Soldiers of the British & Indian Armies During the Victorian Era, 1839-1898.

INDIAN MUTINY 150th ANNIVERSARY: A LEONAUR ORIGINAL

MUTINY: 1857 by *James Humphries*—It is now 150 years since the 'Indian Mutiny' burst like an engulfing flame on the British soldiers, their families and the civilians of the Empire in North East India. The Bengal Native army arose in violent rebellion, and the once peaceful countryside became a battleground as Native sepoys and elements of the Indian population massacred their British masters and defeated them in open battle. As the tide turned, a vengeful army of British and loyal Indian troops repressed the insurgency with a savagery that knew no mercy. It was a time of fear and slaughter. James Humphries has drawn together the voices of those dreadful days for this commemorative book.

LEONAUR

ALSO FROM LEONAUR

AVAILABLE IN SOFTCOVER OR HARDCOVER WITH DUST JACKET

SEPOYS, SIEGE & STORM by *Charles John Griffiths*—The Experiences of a young officer of H.M.'s 61st Regiment at Ferozepore, Delhi ridge and at the fall of Delhi during the Indian mutiny 1857.

THE RECOLLECTIONS OF SKINNER OF SKINNER'S HORSE by *James Skinner*—James Skinner and his 'Yellow Boys' Irregular cavalry in the wars of India between the British, Mahratta, Rajput, Mogul, Sikh & Pindarree Forces.

A CAVALRY OFFICER DURING THE SEPOY REVOLT by *A. R. D. Mackenzie*—Experiences with the 3rd Bengal Light Cavalry, the Guides and Sikh Irregular Cavalry from the outbreak to Delhi and Lucknow.

A NORFOLK SOLDIER IN THE FIRST SIKH WAR by *J. W. Baldwin*—Experiences of a private of H.M. 9th Regiment of Foot in the battles for the Punjab, India 1845-6.

TOMMY ATKINS' WAR STORIES Fourteen first hand accounts from the ranks of the British Army during Queen Victoria's Empire Original & True Battle Stories Recollections of the Indian Mutiny With the 49th in the Crimea With the Guards in Egypt The Charge of the Six Hundred With Wolseley in Ashanti Alma, Inkermann and Magdala With the Gunners at Tel-el-Kebir Russian Guns and Indian Rebels Rough Work in the Crimea In the Maori Rising Facing the Zulus From Sebastopol to Lucknow Sent to Save Gordon On the March to Chitral Tommy by Rudyard Kipling.

THE KHAKEE RESSALAH by *Robert Henry Wallace Dunlop*—Service & adventure with the Meerut volunteer horse during the Indian mutiny 1857-1858.